Western Lands and Waters
XI

THE NATIONAL MINES COMPANY MILL AND ITS ORE-CAR TRACK
Two cars, counterbalanced, could pass on the double-tracked section
in the middle of the run from the upper-level shed near the dumps
to the top level of the mill building. The loaded car
came down, pulling the empty one up.
A corner of the office-residence shows in the foreground.

HIGHGRADE

The Mining Story of
National, Nevada

by

Nancy B. Schreier

THE ARTHUR H. CLARK COMPANY
Glendale, California   1981

LIBRARY OF CONGRESS CATALOG CARD NUMBER 81-66849
ISBN 0-87062-139-4

The word "*Highgrade*" has two meanings.
As a noun, it means "*very rich ore.*"
As a verb, its meaning is "*to steal.*"
It must be used, over and over, to tell
  the story of National and its mines.

Contents

Illustrations

Except where otherwise credited in the captions, the illustrations are from the Pelton collection.

Acknowledgments

If my mother hadn't been fond of the bull terrier that my grandfather, C. W. Buckley, "sent out to the mine," I might never have heard of National. If I hadn't come west and seen Nevada firsthand, my interest in the matter might not have been sharpened. And if I hadn't been living in Los Angeles, I would never have met Mrs. George Pelton, who has been generosity itself in letting me go through her family papers.

I deeply regret that I never met her husband, George Pelton, one of the chief figures in the story of National. By the time I had read through his business correspondence, written while he was absorbed in developing the National Mine, I felt as if he were someone I knew and liked very much. Besides the letters, there were business papers, newspaper clippings, and photographs. In addition to this material, there were personal anecdotes and recollections of his mining days, jotted down at a later date. This collection, which came as a surprise and a delight, has been an extraordinary source of information. My thanks to Mrs. Pelton are without limit, not only for making the records available, but for her endless kindness and warmhearted encouragement.

I met with this same spirit of interest and helpfulness in everyone to whom I talked about National. In Winnemucca, I remember with special affection the hospitality and help I received from the Tony Mendietas, from E. W. "Shorty" Darrah, Harold Cornforth, the Frank Wagners, and J. L. Germain, Humboldt County Recorder and Auditor. They were generous with their time and their stories were wonderful.

In Reno, the late Clara S. Beatty, Marion Welliver, Myrtle Miles, and the staff of the Nevada Historical Society went far beyond the help I asked of them. I shall never forget Mrs. Beatty keeping our two boys motionless and enthralled with tales of her own early days in Nevada. At Reno, home of the University of Nevada and its Getchell Library, I spent days reading microfilm of old newspapers. My thanks to Robert Armstrong and Mary Nichols, who worked wonders in keeping me organized during my various visits to the library, as well as to Dr. Vernon E. Scheid of the Mackay School of Mines, University of Nevada, and to John H. Schilling, Director of the Nevada Bureau of Mines and State Geologist.

In the Los Angeles area, my thanks go to William E. Whelchel, one of the present owners of the National Mine.

I regret that so little material has come from my immediate family. I have remembered a few stories, but neither my mother nor my grandfather saved any photos or written records. I am all the more grateful to my aunt, Mrs. Wilbur C. Cook of Santa Barbara, for keeping her albums with snapshots taken at National. These photos, along with some from Mrs. Pelton's collection, plus a few obtained from Mr. Darrah in Winnemucca and a few that I bought there at Glen's Camera Bar, as well as one sent to me by Maxine Moore of her uncle Ed Smith, pioneer stage line operator, are my whole photographic record of National in its heyday.

Most of all, I must thank my husband, Konrad, and my sons Konrad and Douglas, for making every excursion to Nevada a happy memory. Without their patience and enthusiasm, I would never have seen National, nor written about it.

Introduction

Stories of fabulous mines are older than King Solomon, and rich gold discoveries find as lively an interest today as they ever did. One of the richest, in terms of concentrated wealth, occurred some three generations ago at National, Nevada, where there were mines so rich that thieves broke in to steal the ORE — ore so rich that no miner who worked in the stopes could hope to remain honest. And the wealth was no secret. It was a twentieth century town with telephones, automobiles, a post office, and its own newspaper. It was written up in mining journals, in general newspapers, and was the subject of a government bulletin. To men who had followed the rush to the far-away Yukon and had seen the excitement right there in Nevada at Goldfield, and Wonder, and Rawhide, it sounded like the pot of gold at the end of the rainbow. For those who mined the pockets of gold — and lined their pockets with the gold they mined — the rainbow fulfilled its promise.

Highgrade

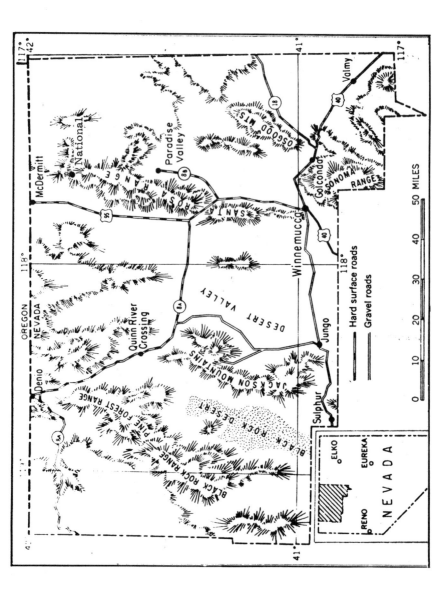

HUMBOLDT COUNTY, NEVADA

1

Automobile Prospecting

The early years of the twentieth century saw a brisk and growing rate of prospecting in Nevada. The great discoveries of silver and gold at Tonopah and Goldfield spurred prospectors to fan out all over the state, and every successful strike made the rest try harder. Jesse Workman was one such prospector, and for him the search led to the Santa Rosa Range of northern Nevada.

Workman had a car, a partner, and a tip from an Indian that there was a certain mountain in the Santa Rosas worth investigating. The location was remote, just a few miles south of the Oregon border and seventy miles north of Winnemucca and the transcontinental railroad. In 1907, the automobile was still something new in the art of prospecting, but a wagon and stagecoach road led north up the Quinn River Valley that paralleled the Santa Rosas to the west and passed within ten miles of the mountain they sought. Workman was a fair country driver as well as a first rate prospector. He and his partner, Lew Davis, chugged through the sagebrush flatlands, following the stage road through sand and dust until they turned eastward to the rounded foothills near the Fort McDermitt Indian Reservation. It was dusty and bumpy and they had to break their trail now, but it certainly was an up-to-date way to travel.

On June 24 of 1907, they found the mountain they were looking for. Charleston Hill, large and rounded, was covered with luxuriant sagebrush and a heavy mantle of decomposed rock which made prospecting difficult. There were few natural outcrops of rocks or vein

material and, because of the mining laws, Workman
needed all his skill and experience in deciding where
to lay the side lines and end lines of his claims. The
right choice in these matters was vital. When he did
decide, Workman posted his location notice for the first
claim, the West Virginia No. 1, while Lew Davis built
the required monuments of rock. Workman signed the
notice for the two men and when the monuments were
finished, used them as a base from which to measure
off additional claims. With a tapeline they measured
600 feet (the legal maximum for the endline of a claim;
for sidelines, the legal maximum is 1500 feet) and built
a second set of monuments. They continued to measure
off their claims and build their corners until all the requirements for legal location were completed. It is a
tribute to Jesse Workman's prospecting skill that of all
of the many claims he located, the first eight on Charleston Hill were the key claims and the productive heart
of the district.

The men were so impressed at coming this far by
car that they named the mining district National, after
Workman's National automobile. Two hills were called
Auto Hill and Radiator Hill, and a map of the National
mining district shows claims named "Fender," "Starter,"
"Brake," "Headlight," "Low Speed," "High Speed," and
"Transmission."

Workman and Davis could not begin to develop all
thirty-four claims on which they filed. They hoped to
sell some of the claims, or lease them to men who would
develop the ground and pay a royalty on any ore that
they mined. The boom camps of Tonopah and Goldfield
had grown rich on leasing, and for the original locators
of claims it seemed to be a marvelous idea. At no cost
to the owner, his claim was developed far beyond the

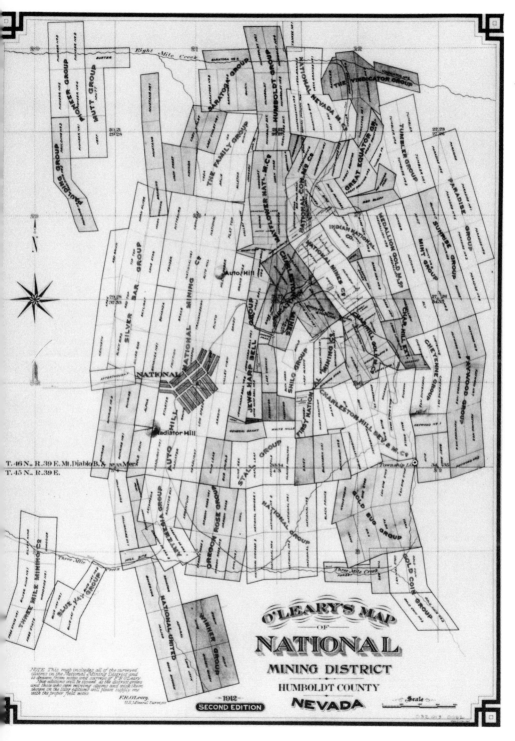

Courtesy of Nevada Bureau of Mines.

A TYPICAL ROAD IN THE DESERT
Automobiles, stages and freight teams experienced difficulties.
Courtesy of Nevada Historical Society.

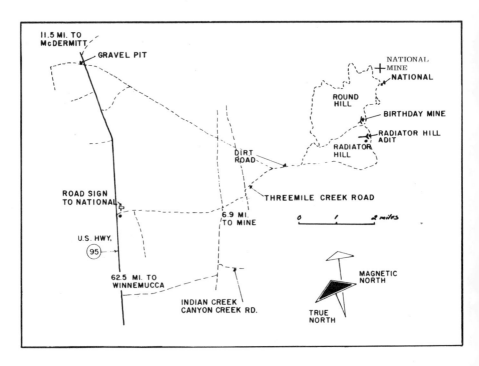

ROADS IN THE NATIONAL MINING DISTRICT
National Mine site is shown by the +.
An excerpt from *National Fallout Shelter Updating Survey, Nevada West.*
State of Nevada Civil Defense, revised 1969. Courtesy of Nevada Bureau of Mines.

Mr. and Mrs. Jesse L. Workman (standing)
From the E. W. Darrah collection.

Early National — A "Ragtown"
Tents, with stoves — and stovepipes — , were home to early comers. These stand
between the town of National and the divide, on the road to the mines.

and upwards..

an.d...Block..22..200.X200..Feet...on..Oh. No....
mining...claim...Consolidated..Aug. 2nd 1905

All ore shipped shall be shipped by and in the name of the First National Bank of Winnemucca, and it to pay from
the proceeds of sale to the Lessor said royalty after deducting the transportation and reduction charges of said
ores; and the balance to be paid by said bank to the Lessee. All milling ores to remain in the custody, care
and control of the Lessee for one year after the termination of this lease, and subject to said royalty, and
after the expiration of said one year, all such ores not worked or reduced, shall be the exclusive property of the
Lessor, his heirs, or assigns.

10. To deliver to said Lessor the said premises with the appurtenances and all improvements in good
order and condition, with all drifts, shafts, tunnels and other passages thoroughly clear of all loose rock and
rubbish and drained, and the mine ready for immediate and continued working, (accidents not arising from
negligence alone excusing) without demand or further notice, on the said 20 day of June
1905........) at noon or at any ......) previously on demand for forfeiture.

11. And finally, that upon violation of any covenant or covenants hereinbefore reserved, the terms of
this lease shall, at the option of the said Lessor, expire, and the same and said premises, with the appurtenances
shall become forfeit to said Lessor, and said Lessor or his agent may thereupon, after demand of possession in
writing, enter upon the said premises and dispossess all persons occupying the same, with or without force and
with or without process of law; or at the option of said Lessor, the said tenant and all persons found in oc-
cupation may be proceeded against as guilty of unlawful detainer. It is agreed between said parties that the Lessor may post notice on said premises that he will not be responsible for labor on said premises or for any
material furnished therefor, etc.

Each and every clause and covenant of this Indenture shall extend to the heirs, executors, administrators
and lawful assigns of all parties hereto.

IN WITNESS WHEREOF, the said parties have hereunto set their hands and seals.

.... (SEAL.)
McMayer..................................... (SEAL.)
G. Donaldson............................. (SEAL.)
C E Stall..................................... (SEAL.)
J B Bradford............................... (SEAL.)
High M Stall............................... (SEAL.)
S. T. .... acting re. for....

"The Star," Winnemucca, Nevada.

Mining Lease.

THIS INDENTURE, made this..2c.day of....June........1908, between J. L. WORKMAN, in person,

the party of the first part, the Lessor, and...R.E Naylor....I B Longbon...

and....G. W. Clark....

the Lessee.

WITNESSETH: That the said Lessor, for and in consideration of the royalties hereinafter reserved, and the
covenants and agreements hereinafter expressed, and by the said Lessee to be kept and performed, hath granted,
demised and let, and by these presents does grant, let and demise unto the said Lessee all the following
described mine and mining property situated on the north fork of Twelve Mile Creek, in Humbolt county,
State of Nevada, together with the appurtenances.

To have and to hold unto the said Lessee for the term of twelve (12) months from date hereof, expiring

at noon on the..20.day of..June.......190 9...... unless sooner forfeited or determined through the

violation of any covenant hereinafter against the said Tenant reserved.

And in consideration of such demise the said Lessee doth covenant and agree with the said Lessor, as
follows, towit:

1. To enter upon said mine or premises and work the same in mine fashion in manner necessary to
good and economical mining, so as to take out the greatest amount of ore possible with due regard to the deve-
clopment and preservation of the same as a workable mine, and to the special covenants hereinafter reserved

2. To work and mine said premises as aforesaid steadily and continuously from the date of this lease,
with at least .... 8 ....persons employed for at least . 15 ...... shifts to the man each calendar
month.

3. To well and sufficiently timber said mine, at all points where proper in accordance with good mining
and to repair all old timbering wherever it may become necessary.

4. To allow said Lessor and his agents from time to time to enter upon and into all parts of said min
for purposes of inspection.

5. To not assign this lease or any interest thereunder, and to not to sublet said premises or any part
thereof without the consent of said lessor, after application made in writing by the said Lessee, and to not
allow any person not in privity with the parties hereto to take or hold possession of said premises or any part
thereof under any pretense whatever.

6. To occupy and hold all cross or parallel ledges, spurs or mineral deposits of any kind which may be
discovered by the said Lessee or any person under him in any manner, by working within or from the demised
ground as the property of said Lessor, with privileges of said Lessee of working the same as parcel of said demised premises.

THE TOWN OF NATIONAL IN THE WINTER OF 1910-11
From the E. W. Darrah collection.

A STAGE TEAM AT THE SNAPP RANCH
A Rebel Creek stage and freight team stop.
Courtesy of Nevada Historical Society.

MR. AND MRS. JOHN E. PELTON
At their "Colonial" apartment in Reno.

RECEIVED at

21 - Q N SM   10 Paid

NX Chicago, Ills, 2/24/9

George S. Pelton,

Winnemucca, Nev.

Our troubles are over leave tonight on overland for

Nevada.

Jno. E. Pelton.

758:PM

MONEY TRANSFERRED BY TELEGRAPH.   **CABLE OFFICE.**

JOHN PELTON'S TELEGRAM FROM CHICAGO
"Our troubles are over."

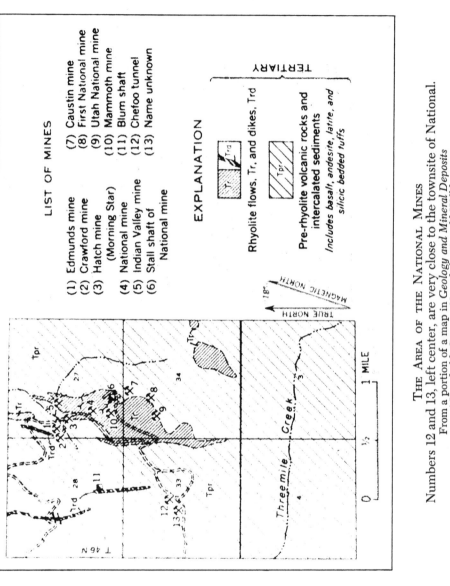

LIST OF MINES

(1) Edmunds mine
(2) Crawford mine
(3) Hatch mine
    (Morning Star)
(4) National mine
(5) Indian Valley mine
(6) Stall shaft of
    National mine
(7) Caustin mine
(8) First National mine
(9) Utah National mine
(10) Mammoth mine
(11) Blum shaft
(12) Chefoo tunnel
(13) Name unknown

EXPLANATION

Rhyolite flows, Tr, and dikes, Trd

Pre-rhyolite volcanic rocks and
intercalated sediments
*Includes basalt, andesite, latite, and
silicic bedded tuffs*

TERTIARY

THE AREA OF THE NATIONAL MINES

Numbers 12 and 13, left center, are very close to the townsite of National.
From a portion of a map in *Geology and Mineral Deposits
of Humboldt County, Nevada* by Ronald Willden.

minimum requirements of the law and, if the lessees found ore, they paid a percentage of its value to the owner. The percentage was open to negotiation and generally varied with the value of the ore recovered. In a typical Nevada lease, not more than ten percent was paid on ores running less than $50 per ton gross value. The percentage rose in steps until it reached "all ores running over one thousand dollars per ton," at which point negotiations got long-drawn-out and impassioned.

In a productive district it isn't hard to find men willing to lease but at National, a newly established district in a remote area with no previous record of ore discoveries, there weren't many people eager to start digging away at Charleston Hill. By the time another spring rolled around, Lew Davis was willing to trade his entire interest in all thirty-four claims to Jesse Workman in exchange for quicksilver claims at Ione, an important mining district in the Reese River valley of central Nevada. S. K. Bradford took a lease on Auto Hill, but he wasn't happy with it, so he teamed with J. E. Kendall to take a lease on the West Virginia claim. Next to them was a lease operated by George and Frank Stall, two brothers from the old gold camp of Marysville, California, by way of the Yukon goldrush. Adjoining these leases to the north was a lease taken by L. O. Donaldson and Roy Gayer.

There was enough activity in the district to justify opening a postoffice on August 7, 1908. Early that spring, some silver float (unattached ore-bearing rock) was found on Charleston Hill, and the lessees began sinking to get to solid ground so that a cross cut could be run both ways to intersect with the hoped-for silver vein. In June 1908, the three leases were consolidated and

were known thereafter as the Combination lease. The
Donaldson and Gayer shaft was farther down the hill
than the other workings, and it was decided to work the
Combination lease through this shaft. Some fifty to
seventy-five feet to the east, a cut was made and found
to contain silver, principally ruby silver. Assays showed
it to run between fifteen and sixty-six ounces per ton
plus a trace of gold. At a depth of forty feet a cross cut
was started going west. When the miners had gone only
fourteen inches, a vein was cut which contained pockets
or nodules of electrum, a natural alloy of about half
silver, half gold. According to one report, the values of
the electrum, containing 49% silver and 51% gold, with
no base metal, ran $78,000 per ton! Five tons of ore
were shipped to the Selby smelter near San Francisco.
The *National Miner*, the local weekly paper, later recalled proudly that, "The first ore shipped . . . aver-
aged $19.90 per pound at Selbys." In 1908, gold was
valued at $20.62 per ounce.

At this point Bradford and Kendall sub-leased their
one-third interest to the Stall Brothers, which left four
men actively occupied in lease development. Meantime,
a man named Sam W. Gundaker secured from Workman
an option on a block of four claims, including the West
Virginia, where the Combination lease was. Gundaker
then interested Harry MacMillan and George Holleran,
mining engineers from Goldfield, in putting up money
for an option on the same property. MacMillan and
Holleran and a third man by the name of Taylor were
the owners of the successful "Engineers' Lease" at Goldfield. In October, the Goldfield men made a payment on
their option and organized the National Mining Co., a
Nevada corporation capitalized at $1,000,000 with shares
at $1.00 par value. Gundaker wound up as president of

the new corporation and manager of the property. Originally, they decide to issue 700,000 of the shares, with 200,000 going to Jesse Workman, 100,000 to be put into the bank to be sold, and the balance divided between Gundaker, Holleran, and MacMillan. Under the terms of their contract, if a second payment on the option was not made by January 1, 1909, the shares of Gundaker, MacMillan and Holleran would be forfeited to Workman.

Besides the option on the block of four claims, consisting of the Fairview, the Fairview No. 1, the Charleston and the West Virginia, Gundaker purchased the Donaldson and Gayer interest in the Combination lease. This left the Stall brothers as operators of the lease which, naturally, became known thereafter as the Stall lease and the Stall shaft.

The year was drawing to a close. Winters are long and hard at an elevation of 6,000 feet in northern Nevada, and if the men had not found promising indications in the ore they had mined that summer, they would not have remained on the treeless slope of Charleston Hill. They were living in tents, getting their water supply from local springs, and boring their shot holes for blasting with hand-held drill steel and hand-driven hammers. The shafts they sunk were each topped with a hand-operated windlass. The rock would occasionally yield fist-size pieces of rich ore, but there was so little of this it barely kept them in beans and blasting powder. All supplies had to be freighted over seventy miles from the railroad at Winnemucca. They could be snowbound for days, and the temperature at National is known to fall to ten and twenty degrees below zero. Still, at least six or seven men were living and working at National when there appeared on the scene an unusual young man.

2

George Pelton Makes an Offer

George Pelton, the newcomer, was scarcely in his early twenties, but he was already experienced in the Nevada mining rushes. He and his father were in Goldfield when it was at its peak in 1906. They were in Wonder in 1907, where young Pelton supervised two prospect developments, one with a shaft and one with a tunnel. Four or five men worked on each property and, in the nine or ten months he was in charge, George got a concentrated course on the ins and outs of survival in a mining camp.

When Wonder began to die, George and two friends piled their bedrolls in a wagon and drove sixty desert miles to Rawhide, just starting its meteoric rise. A property called the Rawhide Royal was available. George took an option on it and his roommate, John Upham, financed the property. Upham was a wealthy young man who found the Rawhide boom so irresistible that he had sold his seat on the New York Stock Exchange and had come west. He and George became good friends.

George's father, John Pelton, also came to Rawhide. As they had done in Wonder, the two worked independently, but they were congenial and John Pelton soon got to know George's roommate.

Late in the summer of 1908 George got a mastoid infection. His father rushed him to Reno, where he became the first man in the state of Nevada ever to survive a mastoid operation! Enforced convalescence gave George a chance to study bookkeeping, always useful, but the Rawhide boom lasted only a short while and by the time he got back there, the Rawhide Royal had run out of ore.

Young George Pelton had done pretty well at Wonder and Rawhide. He liked the life and determined to take his time and look over the mining situation in northern Nevada before considering any other line of business. He based himself in Reno, where his parents had an apartment at the "Colonial," and traveled around to look at a lot of prospects.

None looked too promising, but in September 1908, he started out for one last trip, this time going by train as far as Winnemucca and then north by stagecoach up the Quinn River Valley. After two days of rugged travel, he left the stage at Rebel Creek. The stage stop here was called Snapp's Ranch, where Mr. Snapp, with his wife and daughter, ran a small hotel, a bar, a farm, and rented horses and mules. When George discovered that Pearl Snapp was young and attractive, and that her father was a former mining man, he decided that it would be a good idea to stay at Rebel Creek for a while and investigate the local possibilities.

"This country is all mined out," said Mr. Snapp when George first asked him. "There's nothing left."

That was discouraging, considering how little had been found to begin with, but George was in no hurry to get away from Rebel Creek. When he heard that some men were digging about twenty miles away, back in the hills, he hired a horse and rode over to have a look. Following the wagon tracks as they snaked eastward up Three Mile Creek and then turned north to run past some rounded hills, he eventually found six men working three leases on a steep hillside. This was the National mining district. The owner and discoverer, Jesse Workman, was living in a tent over the hill from the diggings. The men had no hesitation in telling George that the three leases would expire in the spring, and that the option on the property, held by the Goldfield engineers, would expire on December 31, 1908, only a couple of months away.

He examined the shafts that had been sunk some forty-five feet down through unproductive rock. The men told him of the tantalizingly rich pieces of ore that they found from time to time, but had to admit that these were so few and far between that it scarcely paid them to stay there. George was no expert geologist and he was never known as a man who played wild hunches, but for some reason he was strongly attracted by what he saw. He consulted his father, who agreed to put up half the money necessary to pick up the option on the property if and when the Goldfield engineers dropped it.

The last thing George wanted to do at this point was to show any further interest in National. He made Snapp's Ranch his headquarters and made no more trips up Three Mile Creek. In December, the Goldfield engineers spent a night or two at Snapp's Ranch on their way to take a final look at their optioned property. They stayed there again on their return, while waiting for the next stage back to Winnemucca. In casual tones, George asked them about National, and discovered that they intended to abandon their option.

A few days after the New Year, when George was certain that the Goldfield men really had let the option expire, he went up to National to see Jesse Workman. In a horse-drawn wagon (an automobile was not to be considered in midwinter) the two men traveled down to Winnemucca where, on January 10, 1909, they signed a contract. Workman was at this point the sole owner of all of the outstanding shares of the National Mining Co., since the shares held by the option holders had reverted to him when the option was dropped. Consequently, no other individuals had to be consulted when George Pelton made his offer and Workman accepted it.

By the terms of the contract, Jesse Workman agreed to place 200,000 shares of National Mining Co. stock with the First National Bank of Winnemucca. The Peltons were to buy 10,000 shares of this stock immediately, another 90,000 shares by April 10, 1909, and the remaining 100,000 shares by July 10, 1909. The purchase price of all 200,000 was to be ten cents per share. When the 200,000 shares were fully paid for, the Peltons were to receive a bonus of 300,000 additional shares in the company. Since only 700,000 shares had been issued, this would give the Peltons five-sevenths of the outstanding shares. Workman would keep the other 200,000 shares, and the remainder of the original capitalization, 300,000 shares, would continue to be held in the corporate treasury, unissued. When the Peltons had paid for their 200,-000 shares, the 300,000 bonus shares and Workman's remaining 200,000 shares would be pooled for a period of fifteen months unless the board of directors released them earlier. The Peltons would have the privilege of naming three of the five company directors immediately, or as soon as possible. All royalties received from leases were to be credited to Workman and be deducted from the final payment of July 10, and stock issued to the Peltons to the amount of the royalty.

On the same date, January 10, 1909, George S. Pelton wrote a check on the Farmers & Merchants National Bank of Reno, Nevada, for $500.00 payable to J. L. Workman. John E. Pelton put up the other $500.00. Workman placed his stock in the Winnemucca bank and the Peltons were on their way to acquiring the National Mine. At this time its prospects were not particularly encouraging. The only work being done on any of these claims was at the Stall brothers' lease, where George and Frank Stall kept on mining right through the winter months. After the Yukon, mid-January in the Santa Rosas probably didn't seem too bad. At all events, their industry was richly rewarded. Only a few weeks after the Peltons signed the contract that gave them control of the National Mine, the Stall brothers struck a highgrade pocket worth some $3,800.

For the Peltons, the timing of the highgrade discovery on the Stall lease was disastrous. They might have a contract, but Jesse Workman and his lease operators had physical possession of the mine and they absolutely refused to give it up! Siding with Workman and the Stall brothers in this unexpected holdout was Sam W. Gundaker. His stock had reverted to Workman when the Goldfield engineers option was dropped, but he was still president of the National Mining Co. and manager of the mine, and he had purchased the Donaldson and Gayer interests in the Combination lease. They not only refused the Peltons access to the mine, something a mine operator can legally do to a stockholder in Nevada, but they refused to deliver the stock when the money was offered, and they refused to abide by other terms of the contract.

George knew he couldn't do himself any good by staying in the enemy camp so he holed up at the Lafayette Hotel in Winnemucca and kept in touch with a few friends at National, especially P. W. Campbell, mill superintendent and partner in the American Ore Reduction Co. Campbell and his partner, George Hartley, had had unfriendly business relations with Workman in other camps, and Campbell was more than willing to serve the Peltons as their eyes and ears in National. Fearing that the mail might be intercepted, they used plain envelopes, addressed their correspondence in "women's writing," and other dodges. Not all their schemes were successful. A Pinkerton detective was hired to work as a Pelton undercover man at National; he tried to get Gundaker drunk so he would let slip information but, Campbell reported ruefully, "The Pinkerton man passed out first."

It didn't take long to realize that Workman and his pals intended to hang on to the mine as long as they thought they could get away with it. Maybe they hoped the Peltons wouldn't have the determination, or the money, to put up a fight. The Peltons would have to go to court, and they did. On February 13, they sued "Workman et al." for control of the National Mine.

Not only is litigation expensive, but the Peltons knew that while there was a lawsuit going on, royalties from the mine would be impounded, and court action would have to be financed from other sources. Up to now, the National Mine had been mostly George's "baby," but when it became a question of raising funds to carry on a lawsuit, George turned to his father.

John Pelton was willing and able to shoulder the load. He remembered John Upham, George's roommate from Rawhide days, and his financial contacts. In mid-February, John Pelton took the train to Chicago. That was as far as he had to go. On February 24, 1909, he wired George, "OUR TROUBLES ARE OVER LEAVE TONIGHT ON OVERLAND FOR NEVADA."

3

The Chicago Backers

In Chicago, John Pelton had found two men to join him in his fight to gain control of the National Mine. They were Sam Scotten and Joseph Snydacker, partners in a grain brokerage firm on the Chicago Board of Trade as well as in a number of enterprises requiring venture capital. Coal mines, Alaska salmon canneries, timberlands, real estate, gold mines — if it looked promising, they were interested and if they were interested, they told their friends. Sam Scotten was married and the father of several children. Joseph Snydacker was a bachelor with a penchant for actresses. His pince-nez eyeglasses and prissy-prim look were deceiving; he liked pretty women and parties and he had a *white* piano in his Chicago apartment. His manner was cool, prudent, and matter-of-fact, in contrast to Scotten, an excitable Irishman who was prone to repetition in matters of petty detail. The partners agreed to put up the money the Peltons needed in return for stock in the National Mining Co., and made it clear that they would place some of their stock with friends who wanted to invest.

The basic plan of action was worked out before John Pelton boarded the westbound Overland Limited, and things started moving right away. He had hardly gotten back to Winnemucca in late February when he got a telegram: "HAVE WIRED 10 000 YOUR CREDIT FIRST NATIONAL BANK, WINNEMUCCA THROUGH NATIONAL BANK OF REPUBLIC, CHICAGO, UNDERSTAND YOU WILL MAKE TENDER AND IF NECESSARY BEGIN ACTION IMMEDIATELY" (signed) SCOTTEN AND SNYDACKER.

They got his answer on March 1: "19 000 CASH TEN-
DERED WORKMAN AND FIRST NATIONAL TODAY BOTH PAR-
TIES REFUSED ACCEPTANCE, STARTING SUIT FOR STOCK
DELIVERY TODAY, WILL CONSULT YOU BEFORE BRINGING
ACTION FOR POSSESSION OF MINE. LETTER TOMORROW."

Obviously, the Workman faction wasn't going to give
up easily. In Winnemucca, the Peltons retained L. G.
Campbell, an able lawyer, and Sylvester S. Downer, a
former judge who maintained a law office in Reno. On
March 6, George Pelton wired Scotten and Snydacker:
"CAMPBELL ADVISES WE TENDER 19 000 PAYMENT IN FULL
AND START SUIT FOR POSSESSION AT ONCE. WE ARE REFUSED
ACCESS TO COMPANY RECORDS. MINE IS BEING WORKED
UNDER GUARD. NO ONE ALLOWED TO ENTER EXCEPT EM-
PLOYEES. LARGE QUANTITIES OF ORE BEING EXTRACTED
AND STORED IN MINE. NO ORE SINCE CARY SHIPMENT. VALUE
88 THOUSAND TO TON. STILL HOLD 9 000 DRAFT. LOOKS LIKE
BIGGEST MINE IN STATE. WHAT DO YOU ADVISE?"

If the $19,000 had been accepted, under the terms of
the contract, the Peltons would now have five-sevenths
of the stock.

The Peltons faced a double problem: to obtain possession of the stock of the National Mining Co. that was
due them, and to get physical possession of the mine.
Since the ore, a half-and-half mixture of gold and silver
called electrum, was reported to run $50 dollars a POUND
(with gold at $20.62 an ounce), their excitement was
understandable. They sued Workman, the First National
Bank of Winnemucca, and Jerry Sheehan, treasurer of
the corporation, for possession of the 500,000 shares
of stock left by Workman in escrow at the bank.

Workman had slipped across the state line into Oregon to avoid being served. He stayed near McDermitt,
a small town right on the Nevada-Oregon border and
about ten miles from National, and came into town

regularly for his whiskey and mail and to keep in touch with his cronies at the mine. Knowing that it would take a long time to reach him, the Peltons brought suit for a receiver and an injunction to stop work on the lease.

With George Pelton based at the Lafayette Hotel in Winnemucca, John Pelton in Reno, and Scotten and Snydacker in Chicago, the mail was heavy. George Pelton didn't want the Chicago backers to lose interest. In an agony of frustration and desire he wrote to them in late March, "Gundaker showed Hartley figures where the one lease has already produced $38,000 — most since we took our bond (January 10). He showed H. samples and data and H. says from Stalls notes and G. there was $60,000 worth [of sacked] ore in the mine that could go at least $68/lb." He really didn't have to worry. It's hard to lose interest in $68 per lb. ORE.

Over and over they discussed their concern about the difficulty in getting receivership, due to delays granted the defendants. On April 2, George Pelton was finally able to write to Chicago that a suit for receiver and injunction was set for April 10 in Winnemucca. Sam Scotten decided to come west to do what he could to help.

They feared that, while matters were stalled in the courtroom, the richest ore mined by the lessees was being sneaked out of camp with no royalty payment credited to the mine owners. Their misgivings were reinforced by incidents such as that described by Scotten in a letter to George: "I guess your father has told you of finding Gundaker and his chauffeur at Mills City [a train stop south of Winnemucca] last night with three bags of ore — I got off to watch the tail end . . . and your father went to the head end and there found them. They came into the smoker and finally jumped off before the train pulled out." Scotten went on to question, acidly, if Jerry

Sheehan, treasurer of the company, had accounted for all royalties due to the company from the smelter.

On April 20, at a hearing before Judge Pike in Winnemucca, the final hearing on the suit for receiver and injunction was continued, but a stipulation was entered into that gave the Peltons what they wanted: all ore produced at the mine was to be taken by the lessee Frank Stall, and nobody else, to the First National Bank of Winnemucca, and the bank was to dispose of the same in accordance with the terms of the lease, and the plaintiffs (Peltons) might visit and inspect the mine. Just getting this far had taken three months.

Through the spring and summer of 1909, the behavior of the lessees was hard to believe. In March, George Pelton reported to Chicago in bafflement, "It is evident that they are back of the other side in the suit but can't see where they can see themselves clear in coming out in the face of our contract . . . I believe they are melting out the richest ore at the camp and shipping it without our knowledge." Not that the Stalls were all that friendly with the old management. P. W. Campbell, the Peltons' friend in camp, wrote a few weeks later that, "Upon Mr. Stall's arrival in camp last week he remarked to a party who had had some dealings with Workman and Gundaker saying that it was awful to get into partnership with two Son of B - - - so from that I feel that the Stalls are not satisfied."

And how long would the Stall lease run? George wrote to Scotten on May 13, "Stall told me he had an extension [of fifteen months] on his lease and hoped we would not stop him . . . his present term expires in about a month, he says." The Stall brothers claimed that they got the extension on December 27, 1908. The records of the company didn't show it, no copy of such an extension could be found, and there was no direct reference to it on company records. George Pelton therefore claimed when he wrote to P. W. Campbell in July, "The working of ground by them is illegal."

Not surprisingly, there was little public criticism of Workman, Gundaker, and the Stalls by the press and the townsfolk of National. Workman, the discoverer, originally owned all the good claims, which meant he got a share of just about every enterprise in the camp — and a lot of deference. Besides which, people felt that without him, there would BE no National. George and Frank Stall were tall, impressive men and experienced miners. They made a lot of friends and they made a lot of deals. Along with Workman, they had a major interest in many mining properties, both at National and elsewhere. After the rich ore was discovered, they became the chief employers in the district. Their half-brother, William Lehman, was their mine foreman, and the men who worked for them backed them all the way. The Peltons, no matter how impeccable their legal claim to the mine, lacked this political base and economic leverage. They were the late comers, and found the sympathy of the camp to be with the men who had been on the ground from the beginning.

In May, a summons was issued requiring Workman and Sheehan to appear in the matter of the stock transfer, and for Workman, Sheehan, Gundaker and John Doe to account for ore removed and to stop it. In June, L. O. Donaldson, one of the original lessees, wrote to George Pelton that Stall and Gundaker were crushing and panning their highgrade instead of shipping the rock, and Gundaker had taken out a bunch of this highgrade. George replied, "I cannot see how Stall can consent to any such work as reducing the ore at the mine. Both the conditions of his lease and the stipulation entered into in court forbid any such action."

Gundaker was apparently quite open about his activities. The *Silver State News* in Winnemucca reported in July that he had brought in several shipments totalling almost a thousand pounds of $60-$70 ore. The same paper also reported that ore at National was being reduced by pan arrastre, a small, inefficient method useful mainly for highgrade.

While delay after delay in obtaining a court hearing was maddening to the Pelton interests, the pressure was beginning to tell on the men up at National, too. In August, P. W. Campbell wrote that, "Gundaker and Workman are fighting each other. Workman angry and notified Stalls to discontinue operations on the lease." Frank Stall then got angry, Campbell went on, and said he always had "held up" for Workman. Also, there was a strong rumor that the Peltons and Workman had joined against the Stalls and Gundaker.

The Peltons were infuriated when the board of directors of the National Mining Co. declared a one cent per share dividend which Jerry Sheehan, the treasurer, then paid to Workman on the 500,000 shares in escrow in the bank. They secured an injunction restraining any further payment of dividends to Workman and also stopping any transfer of stock or of running the property into debt.

As the lawsuits piled up, work at National slowed down and virtually came to a halt. Finally, on August 31, 1909, the Workman faction and the Pelton group got together and settled the suit out of court. The old officers resigned in a body. The shares of stock were handed over. New officers were named: John Pelton, president and general manager; L. G. Campbell, who had overseen their court battles, vice-president; George Pelton, secretary; Jerry Sheehan, treasurer. Sam C. Scotten was named as the fifth director.

They took possession of the mine the next day.

THE COUNTY OF HUMBOLDT

Geo.S.Pelton,et al.
   VS.
J.L.Workman et al.

TO   S.G.Lamb.   DR.

Mch IIth, To serving Sums & Comp on J Sheehan as the Cashier of First Natl Bank,	$1.50	
Mch IIth To serving Sums & Comp on J.Sheehan.& Mileage,	$1.90	
	$3.40	

April 20th 1909.   *Received Payment*

S. G. Lamb

John. H Pelton, et al   WINNEMUCCA, NEVADA.   Aug 17th, 1909. 190

   VS

National Mng Co et al.

~~THE COUNTY OF HUMBOLDT~~

TO   *S. G. Lamb.*   DR.

Aug 16 Serving Affidavit for Injunction & Order to Show Cause on S. W. Gundaker.	2.00
„ 16 Serving Copy on „ as President	2.00
„ 16 „ „ „ J. Sheehan.	2.00
„ 16 „ „ „ J. L. Workman, M	2.40
„ 17 „ „ „ C. L. Tobin. & M.	2.40
	$10.80

Received payment

S. G. Lamb.

TWO OF THE PELTONS' LAWSUIT EXPENSE BILLS
FROM S. G. LAMB (COUNTY MARSHAL) IN 1909

March 30th 1909 —

Received of John E. Pelton Two Thousand Dollars on account of fees for C. G. Campbell and myself in matters of The National mining Co. and under the employment of Dalton & Amy Decker and John E. & Geo S. Pelton —

S. Downer

National Nev. July 21 1909

Received of John E. Pelton, thirty _____ no/100 Dollars for Mileage and Witness fees,

$ 30.00

P. W. Campbell

RECEIPTS FOR LAWSUIT EXPENSE PAYMENTS MADE BY JOHN PELTON IN 1909

STATMENT of NATIONAL MINING CO. of NEVADA

In account with Jno. E. and Geo. S. Pelton.

(Allowed by act of board of directors, being expenses in
suit as stockholders for receiver and other remedies)

P. W. Campbell for witness fees, serving notice on Stalls,
           mileage etc. $ 99.00

Silver State News publication of summons ------------   35.00

S. C. Lamb for sheriff services--------   ------   29.70

Witnesses Allbrecht, J. Sheehan, Hartley, and Lipman   I5.00

County clerck fees   ------   --------   ----   5.00

P. W. Campbell Team and pay trip to National---------   49.50

Witness fees Tobin and Thos. DeJean----- ----- --- -----   5.00

Pinkerton detective San Francisco ---- ----- ----   ---   I2.45

Amigo (Services and trip to National) ----- ---   ----   80.00

Harroun (Services)   20.00

Noel   (Services)   --- --- --- -- -- -- -- --- --   30.00

Telegrams -- -- -- -- -- -- -- -- -- -- -- -- -- -- -- -- -   70.40

Car fare (Various trips Reno to Winnemucca--Reno to Chicago) 621.00

Hotel expense-- -- -- -- -- -- -- -- -- --. --- -- -- -- -- -- - 271.65

Geo. S. Pelton(Services at $200 per month from Feb.I-I909
                 to Sept Ist. I909)   — -- -- --I400.00

Jno. E. Pelton(Services at $300 per month from Feb. I, I909
                 to Sept. Ist.. I909)-- -- -- -- 2I00.00

                                    $4,843.70.

Total due

(All expenditures above set forth occured betwwen Feb. I, I909
and September Ist. I909)

A STATEMENT OF EXPENSES OF THE PELTONS IN 1909
IN CONNECTION WITH THE SUIT FOR RECEIVER

A VIEW OF THE TOWN IN THE SUMMER OF 1910

Seen from the ridge between the town and the National Mine. Radiator Hill rises behind the town. Pelton's temporary tent encampment is at the right of the road where it curves for no apparent reason. The wooden building between the tents and road was the company car garage.

A Modern Feature of the Town in 1911 was a Row of Telephone Poles Marching up the Unpaved Street. Two-story Building is the National Hotel Courtesy of Nevada Historical Society.

Shaft Collar, Gallows Frame and Engine House on Stall Brothers Lease

It was on this leased ground that the rich electrum was first discovered

SACKS OF HIGHGRADE HEAPED UP ON THE STALL LEASE ARE VIEWED BY TWO MEN
Auto Hill rises in the distance.

A NATIONAL FREIGHT TEAM ARRIVES IN WINNEMUCCA, 1908
Charles Williams was the driver.
Courtesy of Nevada Historical Society.

EIGHT MEN PREPARE TO GUARD $250,000 IN BULLION ON DRIVE TO WINNEMUCCA
George Pelton, wearing a long duster, mounts the lead car,
a 60-horsepower 1910 Oldsmobile. The car at the right is a 1910 Buick.

THE LADIES PREPARE FOR AN AUTOMOBILE JOYRIDE
At the Pelton residence in National. Note the tent fly in the background.

4

Taking Hold

It seemed natural to everyone that George Pelton should be chosen as manager of the National Mine. Besides his ability and experience, he had shown a special brand of dogged determination and it gave him enormous satisfaction to settle in where he had been so long denied access.

George promptly left for National and John Pelton headed for Rawhide to relieve Bob Bolam, superintendent of his properties there, so Bolam could take over as superintendent of the National Mine. At this point the mine was untouched except for the Stall workings. No development work of any kind had been done in the unleased area of the four claims that comprised the property. There was no heavy machinery, no buildings, no proper roads, and not much of a mine. Workman and Gundaker, Sheehan and the Winnemucca bank had all been tigers in the courtroom but when it came to mining, they left it all to the lessees, that is to say, to the Stalls.

One hundred pounds of National ore, worth $60 a pound, was sent for display at the Mining Congress show at Goldfield, Nevada, on September 7, 1909. There was one chunk that weighed thirty-five pounds, and the National display was reported to be the richest hundred pounds of ore shown. Richest, but very likely not the showiest. Natural electrum is a dingy gray, because the silver keeps the gold from appearing yellow.

The publicity helped to stimulate the rush for claims at National. Men swarmed in and soon the whole country was staked with claims for miles around. In all, over two hundred and sixty claims were filed before the excitement was over. Throughout that fall and early winter the entire mining district was vigorously developed and optimism ran high. Early in December the Dwyer lease on Radiator Hill was reported by Winnemucca's *Silver State News* to be "the greatest gold strike in the world." They had uncovered a four foot ledge that ran $215 per ton for the total width, with two inches of $80,000 ore. A new mill was ordered for Radiator Hill, and two shifts were working. Everywhere, men were leasing claims and working them. And promoters were very busy organizing mining companies with the word "National" in the title, and selling shares of stock at ten cents a share.

Highgrading was already in high gear. Not all the miners had the good fortune to work in rich ore but when they did, they took samples home from work. Try as they might, there was no way the mine owners or lessees could stop it. The flow of "easy money" gave the remote camp a reputation for hectic gaiety, lively dance halls, crowded saloons, and an over-supply of rough and criminal characters. The more they roistered, the more the highgraders' reputation spread, and the little settlement quickly became known throughout the mining west as a real "golden opportunity."

National was still mostly a "ragtown" of tents and canvas roofs, with the exception of the National Hotel, a two-story white painted frame structure on the main street, built by Hatch and Blakeslee during the summer of 1909. Workman and Gundaker had laid out the townsite which, like everything they seemed to touch, soon became a matter for litigation. The town was located in a gentle little valley about a mile southwest of the

National mine. A ridge separated the town from the gulch in which the mine was located, and the mine could not be seen from the town itself. The road to the mine was an extension of the main street through National. The buildings were not tents as we think of them today — most of the shelters had wooden floors and wooden framing on which the canvas wall and roof were stretched, with provision made for a stovepipe. There was some frame construction, such as a twenty-by-forty-two-foot restaurant "replacing Mrs. Semcel's tent," but freight rates from Winnemucca, the nearest railhead, made lumber fearfully expensive, and there was almost none to be cut from the Santa Rosas. Naturally, a lumber yard was one of the first businesses that started up. A year after its founding the town consisted mainly of boarding houses, eating places and saloons for the miners, and mostly in tents.

Winter in National was miserable. But George Pelton was too busy to be deterred by mere climate. He did mention to his family, in a letter written in late September of 1909, that there had been three days of storm and heavy fog and the road in front of the house (a tent-cabin) was running water. Cheerfully he added, "We are taking out at least $1,000 per day. I am sure that the production will count up to that ever since we took hold on Sept. 1."

Take hold they did. Some twenty-five miners were put on the payroll and the big freight teams began hauling lumber, timbers, mine rails, ore cars, and large amounts of machinery and tools. The winter was spent in blocking out the mine and in running new tunnels. Counting down from the top of Charleston Hill, the tunnels were numbered 1, 2, 3, 3½, 4, 4½, and 5, and these were driven at intervals down the hillside to intersect the vein at deepening levels. In mid-September the *Silver State News* quoted John Pelton as saying they had discovered ore worth "$1,000 per foot," which spurred on the work. By October, there were thirty men in the National Mine, working on Tunnels 3½ (the Green), and 4 and 5 (the transportation tunnel) and in raising an air passage from Tunnel 3 to an upper level for better ventilation. They undertook to stope out all the ore above Tunnel 3 and to backfill with waste rock to prevent the caving of walls along the old stopes.

Caving was a severe problem in Charleston Hill. The upper levels of the mine ran through material so plastic that "rock" seems a misnomer. Miners in Winnemucca recall that a miner could not drill a full complement of shot holes in the tunnel face and wait until the end of his shift to load them all with dynamite. He had to place the dynamite in each hole as soon as it was drilled. If he waited until he had drilled them all, the weight of the mountain bearing down on the soft talc-like ground would squeeze up the first-drilled holes so that he couldn't slide the dynamite in properly.

When the remaining ore was stoped out of the upper levels, it proved to be as rich as any ever taken from the mine. Unlike the lessees' work, the new management spent money to timber solidly as they went, "permanent and miner-like," commented the local newspaper.

Levels 3½ and 4 carried good streams of water which would help to run the new mill they planned. Tunnel 5 was run through country rock, much firmer than the plastic stuff higher up. This tunnel ran in from the hillside to connect with the bottom of the old Stall shaft; then chutes were put in so ore from the upper workings could be dropped to Level 5 and trammed directly to the crusher to be located at the mouth of Tunnel 5.

A small pan mill was installed near the collar of the Stall shaft to handle the highgrade ore that ran $2,000 per ton or better. Near the mouth of Tunnel 5, a small stamp mill and amalgamator could handle two to four tons every twenty-four hours.

By Thanksgiving of 1909, the population of the town was estimated at 350 to 400 people, and there were about a hundred miners employed in all the various mines. Before Christmas, Ed Smith had his auto stage running from Winnemucca to National. There was a telephone line to McDermitt and Winnemucca. The formal office of the National Mining Co. was moved from distant Goldfield to National, and a sixty-horsepower automobile was ordered for the company, to run between Winnemucca and the mine, as the Peltons wanted their own car to carry the gold to Winnemucca. This was a necessity, as automobile transportation was at a premium all through northern Nevada. In a letter to his mother the previous April, George described an interesting predicament: "The man who was here with the machine is broke down, and the Golconda fellow knew he had the drop on us so is charging us $150. The other man has just as good a car and would have taken us out for $75. Dad got a man offer to come here from Lovelocks for the same money but I thought we had better take the man on the ground."

While George ran the mine, John Pelton ran the business end. In January 1910, he wrote to Scotten, "Our supply and machinery account runs so high because when we took over, the mine was nothing in the way of supplies, lumber, or machinery. I have there now: 3 engines, a nearly complete reduction plant, suction fans, a dynamo (electric generator), 3 blacksmith shops — one on each level — air pipe and track for each level, and 4 ore cars."

In that same month, the company automobile was shipped to Winnemucca by train. The eighty mile drive from there to National was a nightmare for George and Bert Bailey, the driver. The snow was twenty feet deep in places and there was no road to the mine. They finished the trip with mules, George reported, and left the car at the foot of the hill.

On January 20, George wrote a curious letter to his father. "The union was organized last night. We laid all of our men off as the Goldfield bunch has been in here and trying to get control." They had been employing twenty-four men in three shifts, and it would not be hard to replace them, not in the richest mine in the district. The miners were paid between four and four and one-half dollars a day in wages, whether they worked the hard black dyke in Tunnel 5 where they made about ten inches per shift, or the easier ground in Tunnel 4, where they made about four feet in three shifts and had a chance to slip the best highgrade ore in their pockets.

Mail deliveries to National were sporadic. On January 28, George wrote to his father, "Smith came in with his machine today . . . he consents to carry only important letters." He didn't explain how Smith decided which letters were important. Two more members of the Smith family appear in his next letter: "The Smith girl that married the cook spoke to me in Winnemucca and said you told her you would put up a building for them to start a restaurant in National and seemed very anxious to move up here. The way things are now it would not pay by any means.

"The other Smith girl said she wanted to do your letter writing when you were in Winnemucca to keep in practice with her stenography."

Picture the thin smile on John Pelton's face as he replied, "I have not spoken to any of the Smith girls since last spring upon any subject and I have not spoken to any one about a boarding house or any other kind of a house in National, neither do I need any typewriting done . . . that whole Smith story is absolutely without foundation." Clearly he sensed, rightly or no, a cloak-and-dagger flavor to the stenography offer.

Not that John Pelton was above a little undercover work himself. He hired a man named C. C. Thayer to trace some men suspected of highgrading National Mine ore. Thayer followed the trail to the Pacific Northwest. Pelton wrote to Scotten and Snydacker that Thayer had collected enough evidence to convict Gundaker's right hand man. The suspect had worked for the National Mining Co. under Gundaker as a bookkeeper, and his brother was a mine superintendent. A few days before the Pelton interests took over the mine, the ex-bookkeeper left National with two suitcases. "Old Man Chrislem," the teamster, hauled the man and his suitcases to Packard's place, about nine miles from National, where he took the stage for Winnemucca. Chrislem judged that each case weighed about a hundred pounds. Thayer tracked him to Seattle, then Portland, then Boise, where an assay office ran the highgrade into bullion which the man then took to the government assay office and sold to the mint. Gundaker was with him in Seattle. In due time, charges were preferred, but Thayer's evidence did not hold up; the suspected highgrader was discharged after the preliminary hearing.

The weather was terrible that winter. On February 13, 1910 George wrote to John Pelton, "The exhaust broke in the Green [Tunnel 3½] and the men got sick today. Owing to sickness and drunkenness we have only one shift's work done on the Green winze since you left." A couple of days later, he spoke of storms that made it hard to get any work done. "Only four men were working yesterday. The snow has drifted so much it seems like they have to shovel everytime that they take a [mine] car out." By the following week, the snow had drifted half way to his window and he felt he was lucky to get men to work at all. It was "impossible to get a car over the roads. There is water and mud covered with ice all the way and the ice breaks through and there's washouts along the road. Need a team to go anywhere."

5

The Matter of the Stall Lease

The Stall lease continued to be a festering problem to the new management. Not only did they feel that the arrangements on the leased ground were an invitation to highgraders, they felt that the Stalls had no bona fide lease in the first place, and they certainly believed that the Stalls had done all they could to cheat them during the difficult months before the Peltons took over from the Workman, Gundaker group. In October 1909, the National Mining Co. sued to recover the Stall lease. The Stalls were charged wilth illegal possession and an invalid lease, reported the *Silver State News,* and their lease was shut down by order of District Judge John S. Orr. A few days later the litigants entered into a stipulation that all Stall brothers bullion and ore would go to the bank for sale, and the money would be put in escrow. On November 9, Judge Orr enjoined the National Mining Co. to post a $25,000 bond, but the Stall lease remained shut down.

In November an accident occurred at the mine which brought rumors and quick denials. The first reports of the disaster were grossly exaggerated, remarked the *Silver State News,* and the cave-in was an accident, the Stall lease was not dynamited, but fell due to inadequate timbering. Two hundred feet of tunnel collapsed, partially blocking Tunnel 3. Three men crawled out, unhurt. The fact that there *were* rumors of skullduggery, however, pointed up the tensions that existed.

The Stall brothers' mine claim holdings were the sec-

ond largest in the district (after Jesse Workman's) and,
while they could do nothing with their lease at the National Mine as long as the injunction remained in force,
they kept busy. The rest of their old gang didn't give
up either. They had been forced to release the National
Mine and its shares of stock, but that covered just one
small segment of a highly promising area. Three weeks
after the Peltons got control of the National Mine, the
National Consolidated Mining Co. (controlled by the
Stalls) was organized and ready to sell 100,000 shares
at twenty-five cents a share, in minimum lots of 500
shares. The officers were old familiar faces: S. W.
Gundaker, president; George Stall, vice-president; Jerry
Sheehan, secretary-treasurer; Frank Stall, manager. C. L.
Tobin was the fifth director. Their claims touched the
northern boundary of the National Mines ground and
they happily proclaimed that they had $100,000 per ton
ore. No one in National liked to think small.

The Gundaker group was considered a menace to
George Pelton physically, as well as financially. In December, Sam Scotten wrote to John Pelton suggesting
that he hire a couple of bodyguards to protect George
against "Gundaker and his henchmen." Scotten feared
that an attack might be made on George when he traveled between the mine and Winnemucca. A week later
Scotten wrote to George that he was still worried about
"fellows shadowing you in the interest of the Gundaker
gang." George and Bob Bolam, his mine superintendent,
took the threat seriously and George wrote to his father
that one of the men would sleep in his tent with him
because Bolam didn't want him to sleep alone. As a
matter of course, George always wore a pistol on his
daily inspection of the mine. Otherwise, a man armed
with a pick could have forced him to open the door to

the strong room where the bullion and the highgrade ore were stored.

For recreation and exercise, George worked out with a punching bag, and he was a good enough boxer to go a few rounds as a sparring partner for Jack Johnson, the prizefighter, while he was living in Reno. George didn't go looking for trouble, but he did what he could to be ready in case trouble came looking for him.

Sam Gundaker went to Chicago in January 1910, offering to sell his promotion stock in the National Mine. Scotten asked John Pelton if they should buy it at fifty cents a share and pro-rate it to those who bought treasury stock at a dollar a share. Late in February, Scotten wrote, "Our friend (C. W.) Buckley has purchased 31,199 shares of Gundaker's National stock at 35¢ a share. He uses his money and will sell pro-rate to our friends at cost price." Scotten figured that if his friend Buckley didn't get any more stock, the pro-rate for Buckley, the two Peltons, and Scotten and Snydacker would be about 3,000 shares apiece.

Even in midwinter there was a brisk trade in claims and leases in the National district and development work went on feverishly. The Stalls continued to develop another of their leases, working two shifts with a windlass on the Birthdays shaft, and starting a tunnel in the second hill to the north, down Charleston Gulch. Bob Bolam observed, in a newsy letter to John Pelton, that "Hatch and Shea (other lease operators in the district) both working two shifts. Sawyer on Radiator Hill has a vein — mostly sulphides, low values. Halley-Wheeler tunnel on property next to this one is in 182 feet with some antimony and zinc in small seams across the tunnel. This a.m. cut into the rhyolite."

At the National Mine they were still battling snowdrifts and, later, floods. When he could look beyond such immediate problems, George Pelton was trying to negotiate for more lease work at the mine. In spite of difficulties with the Stall brothers, leases were considered to be a good thing. Two men by the name of Atkinson and Shea wanted an eighteen-month lease. George was willing, provided that they drove a tunnel, starting in the Fairview and crosscutting the entire width of the Charleston claim. Scotten and Snydacker put in a good word for J. H. Causton and D. H. Jarvis. Jarvis, with whom they had had dealings before, was a good man, they said, and if George wanted to lease, Jarvis should be the man to get it. George was willing, and Causton and Jarvis got their lease just south of the Stall brothers' lease on the West Virginia claim.

In February they struck ore in Tunnel 3, about seven pounds of which was nearly solid gold. John Pelton sent Sam Scotten and Joseph Snydacker each a cut and polished sample of ore that ran about $100 a pound. The Chicago men sent them to a refiner to have the metal extracted and made into pocket pieces. The Peltons themselves showed more imagination. They had some of the richest ore polished cabochon-style and mounted as jewelry.

In April 1910, the matter of the Stall lease was resolved. "I am fully satisfied now," wrote John Pelton to Chicago, "since the settlement of our difficulties that we will get a square deal from the Stall boys." The new agreement gave a royalty of 40% to the National Mining Co. on all ores running over one thousand dollars per ton. Better still, it gave the company increased control. George had steadfastly maintained that, in return for extending the lease, he needed control and supervision of change rooms where the miners kept their working clothes. This wouldn't stop the highgrading altogether, but he considered the arrangement an absolute necessity. Now there would be only one entrance and exit, with a change room, and the company would have control of all the ore as soon as it reached the shaft collar. The Stalls could use the tunnels *only* in their lease block; they couldn't take anyone into the mine without the Peltons' consent; George Pelton could fire any of the Stalls' men that he wanted to (in peak season they had about one hundred on the payroll); the company had right-of-way through the lease block; and the Stalls must make full settlement and turn over all receipts from smelters, strictly accounting for all ore taken from the mine by them. The new lease ran one year and twenty days. If the ground had been clear of all disputes, said George, he would not have leased on any such royalty.

6

Spring, 1910

It was obvious that they needed a bigger mill. The National Mine was producing ore that ran from $50 to $1500 per ton *in addition to the highgrade,* and it would save $40 to $50 per ton in freight rates to mill it on the spot. John Pelton went to San Francisco to look at a Lane slow-speed mill. He considered it the best amalgamator on the market and figured it should cost about $4,000, while an old-fashioned stamp mill of equal capacity would cost around $15,000 and would be less efficient.

C. C. Thayer, who earlier had been on the trail of some suspected highgraders, came to National. Apparently Thayer was assumed to be a man of many talents, as his job at the mine was to set up the new mill when the machinery arrived. George Pelton soon realized that this wasn't going to work. In a letter to his family on April 17, George wrote that Thayer was no good at setting up a mill, and that they should have a Lane mill man install it. The following day he was more emphatic. "Get a good factory man to get the mill in running shape. Thayer tries hard, but his 'fort' is not installing mills . . . He hasn't a mechanical idea in his head." George himself was not so afflicted. "I had to go down to Snapp's lower ranch," he wrote, "and get a pulley off a thrashing machine which we will fix for our pan [for the little highgrade mill] . . . it will give us about 20 revolutions."

The whole Pelton family proposed to move to National

for the summer, and George got busy providing living quarters. His letters home are full of these preparations. April 20: "I will get things ready for Mama to come up. I'm going to move the big tent back for the boys [himself and his younger brother] and fix a couple of small ones up there with good flooring and a glass door and window and a little porch." April 23: "Couldn't get a big tent but have one 12x24 to make a 10x12 kitchen and a 12x14 dining room . . . You can send a cook up any time." April 24: "I think that I will have things in shape by the 28th for the cook so you had better send one a couple of days before you come as then Mother and you will not have to go to these dirty restruants [*sic*] . . . I can have things in good shape provided you send the cook in plenty of time." April 28: "May will be the nicest month for Mother to come up. I will have tables and everything made. We will be in good shape before you get here . . . I hope you send [a cook] soon." George obviously was getting pretty worked up at the prospect of decent home cooking.

Mr. and Mrs. Pelton, their daughter, Leonora ("Leo"), and their younger son, Herbert, came up in May to find the tents all ready, each shaded by a big white canvas "fly" over it.

More visitors to National in the same month included Sam Scotten and Charles W. Buckley of Chicago. Buckley, one of the earliest Chicago investors, was described by the mining camp's weekly paper, the *National Miner*, as "the philosopher, the wit with some humor and the real story teller of this millionaire bunch. His friends relate of him that he always tells the truth, the straightforward truth, which makes him a humorist . . . Buckley is one of the big men in Chicago . . . his grain elevators are dotted all over the northwest and as master of the

Seated at right, John Pelton holds his ever-present cigar.
His trusted superintendent, Robert Bolam, sits next to him.
Behind Pelton is L.G. Campbell, the attorney who saw the Peltons through
the difficult days of litigation. The fourth man is unidentified.

NATIONAL MINE VISITORS GATHER ON THE OFFICE STEPS
John Pelton, far left center, stands next to Sam C. Scotten. C.W. Buckley
leans to pat his dogs. Behind him in a light jacket is superintendent
Robert Bolam. George Pelton is at front right. Others are unidentified.

AT THE MILL BUILDING UNDER CONSTRUCTION IN 1910
John Pelton stands at center with Chicago backers Sam C. Scotten (left),
Joseph Snydacker (right), and Charles W. Buckley (seated).
Courtesy of the collection of Mrs. W. C. Cook.

THE SUPERINTENDENT'S OFFICE AND PELTON RESIDENCE, ALL UNDER ONE ROOF
Built in 1910 across the gulch from the National Mine and its mill.

THE NATIONAL MINING COMPANY'S INITIAL RUN
These eight bricks, with the company's initials, represent slightly less
than 1,000 ounces of gold. They were put on display in the window
of the Farmers and Merchants Bank in Winnemucca in April 1910.

CHARLESTON HILL
BEHIND TUNNEL No. 5
Before the mill was built.

THE MILL UNDER CONSTRUCTION IN 1910

Construction Days at the Site of the Mill

GUARD TOWER AND SEARCHLIGHT
On the hill opposite the mill.
In the tower, day and night, was
a guard armed with a rifle.

THE MILL ON A WINTER NIGHT
It stands illuminated by the
searchlight from the opposite hill.
A second searchlight was
fixed to the mill's "topknot."

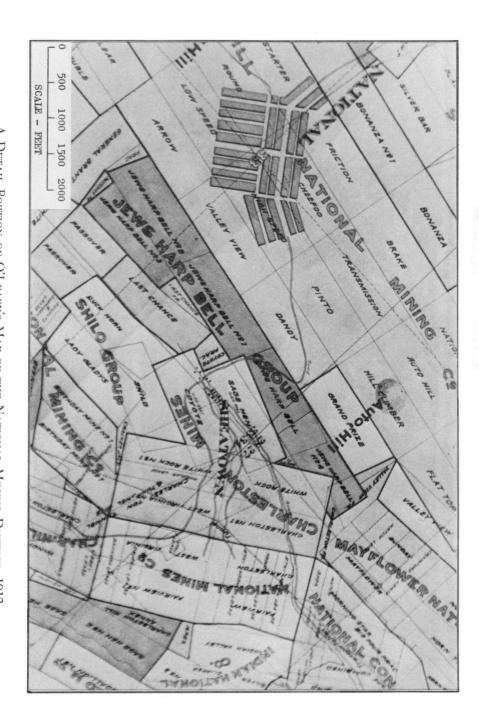

A Detail Portion of O'Leary's Map of the National Mining District, 1912 Showing the townsite of National and the road past Sheatown to the National Mines Company ground. The outline of a square on the West Virginia claim of National Mines Co. marks the site where gold was discovered.

CHARLESTON HILL ABOVE THE MILL
Showing the outward form of the mine levels.

wheat market in Chicago he stands as one of the potential figures in the country."

The Chicago men timed their visit well. They were on the spot when the miners hit ore in Tunnel 4 "better than the Stall lease."

The little mill — and Thayer — were kept busy. Running batches of 250 pounds of ore at a time, Thayer worked it into amalgam which was then retorted to remove the mercury and leave sponge bullion. George assigned a guard to protect Thayer and the amalgam, and John Pelton reported to Scotten and Snydacker that they would soon be able to market their own bullion.

His promise came true before the end of April. Less than eight months after the Peltons and their Chicago backers took over, the company proudly set up a display in the window of the Farmers and Merchants Bank in Winnemucca. Eight gold bars, heaped up, had a handwritten sign perched on top:

INITIAL RUN
NATIONAL MINING CO. OF NEVADA
Returns from less than *one-half* a *Ton* of *Ore*
Value about $20,000.00
Exact *weight* of *Ore* 959 lbs.

The sign was only a little optimistic. When John Pelton took the bullion to the Carson City mint, the return came to $17,315.54.

All that spring and summer, the small gasoline-powered pan mill at the National Mine kept busy working highgrade ore, and a steady flow of bullion shipments began in June. The first one from the Stall lease was worth $92,160. Later that month, 709 pounds worth $176,600 from the Stall lease and 174 pounds from the National Mine were shipped in two automobiles with eight armed guards led by the Peltons and their driver, Charles Haeffner. July saw more armed convoys, with the mine shipping out some $100,000 per week.

With that kind of production, the directors were able to declare their first dividend, of five cents a share, on July 21. Less than two weeks later, the second five-cent dividend was declared. They hoped to be able to continue at that rate every month!

The regular milling grade ore was stockpiled during the early part of the summer. Supplies and machinery for the new Lane mill arrived in June, and construction began right away. The mill was located on the slope of Charleston Hill with the highest part of the equipment, the crusher, at the mouth of Tunnel 5. This permitted the ore to be trammed directly from mine to mill, and each subsequent milling operation took place farther downhill than the one before it, allowing gravity to help move the material. A similar arrangement within the mine allowed ore from the upper workings to be dropped through chutes to Level 5, where it could be transported directly to the crusher.

The thirty-ton Lane mill was a logical choice for such free-milling ore. Briefly, in it the ore was finely crushed by four steel rollers, each seven feet in diameter. The particles of precious metal in the crushed ore were combined with quicksilver on amalgamating plates. The quicksilver was then driven off by heating, retorting, leaving the silver and gold which was then cast into bullion bars. The amalgamation process didn't get all the gold so, after leaving the mill, the tailings passed over two vibrating Johnson tilt tables which saved a big percentage of sulphides in the ore. The concentrates made up about five per cent of the ore milled, and brought upwards of $1,000 to $5,000 per ton of concentrates at the Selby smelter. Although clean-ups every two or three days reduced the nominal 24-ton per day capacity of the mill, the big freight teams started hauling concentrates to Winnemucca every few days for shipment to the smelter.

In addition to the new mill, other construction was going up at the National Mine. The Pelton family were cheerful about living in a collection of tents during the summer, but such an encampment could be only a temporary arrangement. A permanent building to house the mine offices and the Pelton family *and cook* was put up across the gulch from Charleston Hill. In design it was an elongated frame bungalow, simple, functional, and surprisingly attractive. Near the entrance to Tunnel 4, a little house was built for superintendent Bolam, his wife, and child. The new mill itself had more style than was usual with such mine buildings, with a conical "top-knot" built above the crusher, the highest part of the structure.

A marine lighting plant with a 20-horsepower gasoline engine was installed to run the dynamo and the air compressor. It used fifty gallons of gasoline every day. To discourage thieves and bandits, a steel-lined tower was built across the gulch from the mill and armed guards stood watch twenty-four hours a day. On top of this watch tower a powerful searchlight was mounted so that it illuminated the mill buildings at night, as did a second bright light located in the "top-knot."

Domestic water was no problem for the residence-office building. A small tunnel was run to a spring several hundred feet up the hill. No pump was needed. As George recalled, "There was a little dyke, or dam, and a pipe running through it so the water would come clear and free from surface leaves and so forth."

Water for the mill came from the mine, and was a really serious problem. As in many mines, the blasting had caused little fractures in the rock, through which water seeped and dripped all the time. In the National Mine, the mineral sulphides made a lot of sulphuric acid in the water, and the acid ate everything. To keep the mine rails and their tools from rusting away, the men buried just about everything in lime. It was hard on the miners, too. No matter what they wore, the acid bit their clothing so it was soon in rags, and their skin got raw and sore.

As an added aggravation, the water flow to the mill was uneven, depending on the time of year and the weather, and the acid water reacted with the ore — and the machinery — and made the milling process more difficult. The sources of water were so limited, though, that they took the mine water and were glad to get it.

7

Highgraders

National's reputation for wholesale highgrading eclipsed even that of Goldfield. The nature of the ore, with fabulously rich pockets and lenses of almost pure electrum, made it easy for a miner to hold out small pieces for himself. A miner rationalized that with all that wonderful stuff going to the company, surely it was no great sin to keep a little of the best for himself — "family ore" he called it with a smile, as distinguished from "company ore." Some called it "jewelry ore," others "blossom rock"; the mine owners called it grand theft.

The National Mining Co., having by far the richest mine, had the most to lose, and naturally did everything it could to suppress the practice, but it was an uphill fight. After all, said the Peltons, the Stall-Workman-Gundaker faction had set a lax example from the first. In any event, with ore so rich, the temptation to steal was overpowering. The management used armed guards, search-lights, change-rooms, and every precaution they could think of. Even so, the Reno *Journal* reported thefts of highgrade at the National Mine and leases. In June, two masked bandits held up three miners on the night shift as they worked in a stope above Tunnel 3½. The bandits forced shift boss Bob Smith to take them to the sacked ore and forced auto stage driver Ed Smith to go with them. They carried off a one hundred pound sack of highgrade ore valued at $3,000. Both Smiths were released unharmed.

It is a remarkable fact that the heavily-armed convoys

of bullion were never attacked on their way to Winnemucca. All the trouble occurred closer to the source.

In desperation, a private detective was called in.
Charles Siringo, a Pinkerton operative for many years,
gives a lively account of his undercover work at National
in the first edition only, 1927, of his book *Riata and Spurs*.
In Siringo's account, he met Scotten, Snydacker, C. W.
Buckley and John Pelton by pre-arrangement at a railroad "side station on the desert west of Ogden, Utah."
Siringo was instructed to find out how the highgraders
got the ore out of the mine and how they later disposed
of it. He thereupon went to National under an assumed
name and became a pal of the "Gumshoe Kid," leader
of the highgraders. He saw the kind of body harness the
men wore beneath their clothes so they could walk out
of the mine, after their shift was done, carrying as much
as five or six pounds of ore without showing telltale
bulges. The lunch buckets of men going off shift were
inspected routinely, as were bulging pockets, so the
miners had to be cagey. Since there were more than a
hundred men working in the rich ore, and since the ore
was running from thirty to ninety dollars *a pound*, the
highgrading losses were truly staggering. Siringo cites
the case of a young miner who lived next door to him
who had managed to steal $7,000 worth of ore in a few
months. The young man saw his shift boss stealing
as well as the miners working under him, and he knew
that the shift boss *knew* they were stealing. So, after
fighting temptation for three or four weeks, he began
highgrading along with the rest of them.

There were at least ten assay offices in National, far
too many for the size of the mining district under or-
dinary circumstances. Buyers at some of these offices
would pay for stolen ore at half its value or less and

ship it out in a number of ways. A favorite method of the "Gumshoe Kid" was to hide it under the load of a produce wagon. Most of the stolen ore was processed at one of the many different "dumps," secret small portable mills hidden away in the mountains around National, where the ore was melted and cast into bars. Siringo thought there were six of these "dumps"; other reports say "over a dozen, plus one at Willow Creek."

Gangs of ore thieves fought among themselves, and occasionally gunned each other down. At least once some pack mules, loaded with stolen highgrade, were "spooked" and bucked off their loads, piecemeal, as they took off across the countryside. The treasure hunters who followed were never sure that they found it all.

Siringo seconded George Pelton's idea of a change room with armed guards present. These rooms, where the men stripped off their working clothes — completely — under the watchful eye of company guards, had been introduced at the Goldfield mines some half dozen years earlier, for the same reason. In time, the change room was installed and the highgrading went down. So, says Siringo, did the number of dance halls in town — from six to two. Presumably, a few assay offices went out of business as well. Scotten later told Siringo that more than a million dollars had been stolen from the company by highgraders, and that Siringo had managed to save another half a million from going the same way.

"Our friend Buckley" from Chicago couldn't help much to stop the highgrading, but he offered his own solution to the bandit problem. At his home in suburban Highland Park, Illinois, he had a pit bull terrier named Taylor. Taylor had two sons, Tip and Imp, both as black as their sire was white. All three were convinced that their mission in life was to attack the newspaper delivery boy. The paper boy, for his part, wasn't too smart. He had been told to leave the paper at the entrance to the driveway, and never to ride his bicycle up to the house. One day he didn't follow instructions. After paying the doctor bills, and for some new trousers, a new bicycle, and for the paper boy's wounded feelings, Buckley decided that Taylor and Tip and Imp were wasting their talents in Highland Park. The dogs were shipped out to National where they were put to work guarding the strongroom of the National Mine. They earned their keep, the story goes, for after the dogs were set on guard, the strongroom was never broken into again.

When it came to prosecuting a man for highgrading, John Pelton was reluctant. Besides, if a man were hauled into court on that charge, the kind hearted juries of National would surely find him innocent, regardless of the evidence, so it seemed a waste of time and money.

George reflected this attitude earlier when C. C. Thayer ran a suspected ore thief to ground in Seattle. To his father he wrote, "Drop the highgrade law suit — now the mine is making money, so what is spent on law suits can't be declared as dividends." Once George wrote that he "found out that a fellow had 30lb of ore that would run about $1,000 per ton in his tent. I went down when he was away and looked at it and it was from our mine so I took it out of his tent. I guess he will be surprised when he comes home. He did not get much value on ore but his intentions were allright." On another occasion, he confronted a mill-man who was stealing. The man offered to return all the money he had taken, plus the company's expenses in tracking him down. George accepted his offer, and there the matter ended.

They still tell a story in Winnemucca about a man who worked as a teamster in National. One day "John

Teamster" was called to the National Mine office to drive John Pelton and his daughter, Leonora, somewhere. As he waited in front of the office, which was also where the Pelton family lived, "John T." noticed some particularly fine pieces of specimen ore piled in a heap underneath the porch of the building. When he thought no one was looking, he picked up a small piece and tossed it on the ground in front of his wagon team. When John Pelton and Leonora came out and climbed in the wagon, Pelton glanced under the porch and remarked to his daughter, "It looks like one of our specimen pieces is missing. We must have been highgraded."

"John T." paid close attention to his horses. And never said a word.

While the Peltons were getting settled in the wagon, "John T." casually picked up the piece he had tossed near the team, making sure that Pelton saw him "find" it in the road. Putting it in his trouser pocket, he climbed up to the driver's seat and away they went. A little later he felt a sneeze coming on and hastily pulled out his handkerchief. As he did so, the ore sample fell out of his pocket. For a moment it just lay on the floorboard, between his boot and John Pelton's, who was sitting next to him.

"John T." sat stiff and still, gazed at his horses' ears, and never said a word.

"Look at this," said Pelton to Leonora, picking up the ore sample. This looks a lot like the piece that's missing from the specimen collection. Look closely. You'll never see ore richer than this." He turned to the driver and handed him the sample. After all, he had seen him "find" it in the road.

"Here, John," he deadpanned, "I believe this fell out of your pocket. Put it back in so it won't fall out again."

After that, "John T." would have died for John Pelton, but he just nodded, and pocketed the ore. And never said a word.

The change rooms put an end to the use of the body harness as a way to get highgrade out of the mine, but the ingenious miners didn't give up trying. One carried a large pocket watch without any works, but this was rather minor. Only an ounce or so would fit in anything smaller than an alarm clock. A more common practice was to tram a mine car of rich ore to the waste heap and dump it *carefully* where the thief could come back later to recover the ore. This was one of the main reasons for the searchlight towers.

According to Frank Wagner, who worked in the National Mine during the Depression years, one ruse he heard about was to take small bags the size of Bull Durham tobacco sacks or a little bigger into the mine, fill them with pieces of highgrade, and hang the bags on cup-hooks fixed to the underside of the seats in the portable privy used in the mine. They could be retrieved later when the privy was taken out of the mine to be dumped.

There was another method he recalled hearing the old-timers describe. The problem facing the highgraders was a steel door kept locked between the mine tunnel and the change room where a company guard was stationed. Any highgrade had to be left behind the door inside the mine before the men changed, or it would be confiscated. The layout: a small, indescribably filthy drainage ditch ran down one side of the tunnel and flowed out in a little channel beneath the door. The solution: the men put their highgrade into tobacco sacks, or tobacco tins, or wrapped it in pieces of rag that they brought with them. When they came off shift the tins, sacks, and rag bundles were surreptitiously dropped in the foul muck of the drainage channel just inside the door. Then the men came out, changed, and walked away, exuding innocence. After the change room guard had left, the men returned with long iron rods, each with some kind of hook on the end. They slipped the rods under the door and along the channel, "fishing" for the prizes they had dropped a few hours before. Frequently there were bitter quarrels and fights over which Prince Albert can or Bull Durham bag was whose.

"Used to get pretty lively, I hear," laughed Wagner. The highgrading never did stop, was his conclusion. In the National Mine, indeed, in all the mines of the National district, any ore worth mining was worth stealing.

Little has been said about outright bribery of the change room guards. It was sometimes done, surely, but from the number of other methods used to get the ore safely away from the mine, there appears to have been almost a sporting element to the practice of highgrading.

How much *was* highgraded? Well, one of the National Mine managers estimated that some seven to eight millions in ore were stolen, about the same amount that was legitimately produced in the district. Even the town's own newspaper, the *National Miner*, thought this estimate was correct.

If it seems inconceivable that highgrading could account for half of all the values taken from the mines of National, it must be remembered that the mine owners were the only people who were *not* in favor of it. Every other mortal in the district was exposed to temptation, and the richness of the ore made it possible. For all their efforts, there seemed to be no way the mine operators could stop the steady drain.

8

Mining Camp Life

When the National Mine seemed well set on a profit-making basis and the highgraders' pockets were heavy with loot, the town of National began to show characteristics of a permanent settlement.

The two-story hotel, the postal service and a phone line, a lumber yard and a general store were already established. During the summer of 1910 came more signs that the town was there to stay. For one thing, Vol. I, No. 1 of the *National Miner*, founded by Roy Harris, came out on July 22. The population rose to almost one thousand people, and a school district was established. When the school trustees found to their dismay that neither Humboldt County nor the state of Nevada would advance money for a school building, a private subscription was taken up, raising $400 in ten days. Thereupon, Morris Tracey and J. B. Lamb, school trustees, bought lumber and put up the 18-by-30-foot schoolhouse, doing most of the work themselves. A Miss Langwith was hired as the teacher and over forty children were enrolled during the first year.

There was talk of putting up a Catholic church. A priest traveled to McDermitt and National to hold services, but no church was ever built. Just as there was no church, there was no cemetery, although there was an undertaker. The dead were taken to Winnemucca for burial. So much for the enduring legend of a "boot hill" for every mining camp.

The townsite itself was a matter for litigation and uncertainty. According to Jesse Workman, "At the be-

ginning of 1908 . . . people commenced to come in and
a townsite was started on some of my claims according
to prior customs and law in this state. In 1910 others
began to scheme to annul my right to the ground and
rob me and profit thereby, by establishing a government
townsite." He was referring to the fact that in April
1910, a certain W. S. Palmer, who claimed he was a
mining inspector for the Government Land Office in
San Francisco, started a rumor that the townsite would
be declared open by the government. This caused an
immediate rush by forty or fifty squatters for town lots,
and by nightfall there was some kind of building started
on every vacant lot in town. The stores were sold out of
tents and lumber, and tents already put up in outlying
locations were hastily moved to better spots. Workman
defended his claims and asked the Department of the
Interior for a ruling. Whoever owned the land would
indeed "profit thereby." Tom Defenbaugh, owner of a
general merchandise store, paid $2500 for a lot across
from the National Hotel.

Besides National, which was basically a double row
of tents and houses about a mile long, there were other
settlements nearby. "Sheatown" started with a rooming
house built near a spring on the west side of Charleston
Hill. The mine-owned buildings near the mill, at the
foot of Charleston Hill, were where the managers lived.
And there were a few buildings up on the Stall lease
near the top of Charleston Hill. At each location, reported Winnemucca's *Humboldt Star,* was found an "abun-
dance of cool refreshing spring water."

Prices were high. Fourteen-horse and sixteen-mule
freighting teams brought a steady stream of food and
merchandise. A new wagon road was built with *only*
a 7% grade, considered a real improvement over the
original one that followed the creek bed into camp.

Freight rates, which had run $30 a ton when there were only a couple of teams on the route, went down as the number of teams increased. At one point there were some thirty freighting teams making the round trip, but the long haul from Winnemucca added painfully to the cost of living. The going rate for board — as of June 1910 — was $37.50 per month, or $1.25 per day. A bunk in the same house where $37.50 was charged for board was $5.00 per month. As for food itself, butter was 45¢ a pound, eggs 50¢ a dozen, bacon 44¢ a pound. These prices must be considered in the light of their own time when dollars were much harder to come by. They were four or five times the "civilized" average.

The whole camp found it difficult to accept the fact that the National Mine was the only mine making any money. It wasn't because it had such a big piece of Charleston Hill. The National Mine comprised a modest four claims. But while those four claims were producing millions, the rest of the operators were sinking more into their mines than they were getting back. They did find many pockets of rich ore, but never enough to turn the red ink to black.

Climate and remote location combined to give a strongly "seasonal" quality to life at National. High freight costs made housing expensive, with the result that most of the people kept on living in tents. Living in a tent all winter was such a miserable prospect that many of the miners, the ones who weren't working in the highgrade, "went down below" in the fall, and came back in the spring. It was largely a bachelor existence at the camp whether a man was married or not. Most of the miners did not bring their families with them — if they had families — when they made the annual trek to Charleston Gulch.

Travel between National and Winnemucca was not

a casual matter. A miner could journey by auto stage, paying a $12 fare and 3¢ a pound for baggage over twenty five pounds. He could save a little money by taking a four-horse stage, stopping at "frequent stations at which first-class meals can be obtained at reasonable rates." Either way, the trip took a full day or more.

Commercial enterprises at National reflected the life-style of the town and their advertisements in the *Miner* showed up the chief concerns of the population, both temporary and permanent. A lot of space was devoted to freighting and stage lines, both horse-drawn and auto-mobile. The Winnemucca-National Auto Line, Fred Robinson, proprietor, advertised "THREE BIG CARS . . . WE MEET ALL TRAINS . . . Cars Chartered for All Points in Northern Nevada." Frey's Big Mule Outfits special-ized in hauling heavy freight. Summerfield-Pearce Co., Inc., promised they would deliver coal as well as "Both Light and Heavy Hauling." Haviland and Hoskins' Winnemucca-National Stage Line posted a schedule of daily departures of four-horse coaches. The Thomas-Simplex Auto Line advertised "first class machines. Care-ful and experienced drivers — quick time." Speedy ser-vice was equally promised by "The Old and Reliable AUTO LINE, E. A. Smith, Mgr.," who urged the public to "reserve seats with us."

The Western Pacific and Union Pacific railroads ad-vertised their up-to-date "electric lighted" cross-country trains. The National Bottling Works (Soft Drinks of Best Make and all Kinds supplied to the Public) was pleased to mention its name and, of course, saloons, restaurants, cafes, hotels and lodging houses all took space. Lawyers and doctors advertised, but they were almost all located in Winnemucca. The same was true of drug stores, banks, and stockbrokers. National had a shoemaker and one could buy clothing at the general mercantile stores,

A Six-horse Stage Leaves for National
In front of the Lafayette Hotel in Winnemucca.
Courtesy of Nevada Historical Society.

"The Old Reliable Auto Line, E. A. Smith, Mgr."
It carried passengers and mail between Winnemucca and National.
In a 1908 Pope-Toledo the driver and passengers pose
in front of the El Dorado Hotel in Winnemucca.
Courtesy of Nevada Historical Society.

GEORGE PELTON HUGS TIP, WHILE HIS SISTER, LEONORA, HOLDS TAYLOR
A director, C.W. Buckley of Highland Park, Illinois,
sent the dogs to National to guard the highgrade.

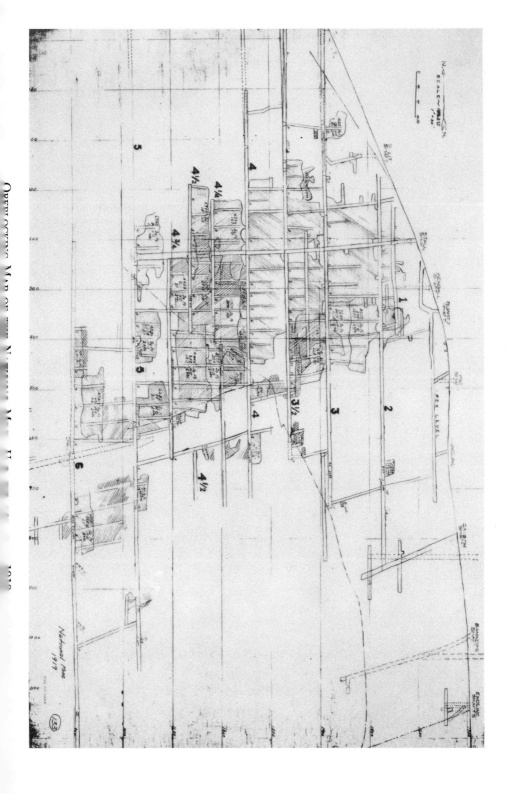

OPERATIONS MAP OF THE NIPISSING MINE, No. II

National Mine
1919

133

GEORGE S. PELTON
Standing at the machine shop near Tunnel No. 5

though variety was limited. Gayer and Donaldson's bath house and laundry filled a need, as did Daniely and Walker's blacksmithing and general repair work, the National Livery stable, and various mining engineers, surveyors, assayers, and chemists. The publisher of the *National Miner* doubled as notary public. Dr. Joseph B. Hardy, local physician, was the official health officer.

Way stations also took advertising space. The Snapp Station at Rebel Creek termed itself "A Place for Traveler and Beast. Good Care for all Such. Feeding and Stabling for Freighters ample." L. C. Peterman of Amos, thirty-two miles north of Winnemucca, promised "First Class Service and Accomodations" for his Castle Creek Stage and Freight Station, as well as for the station he kept at Cane Springs.

As might be expected, adventurers of every stripe were drawn to the lure of the highgrade, and the freewheeling camp attracted a sizeable number of rough and unsavory characters. The town had a jail, a constable, and a deputy sheriff, who depended on Winnemucca for law courts. S. G. Lamb, the county marshal, wore a gold star made from ore given him by the Stall brothers. Violence came early to National, sometimes in the form of gunplay, but not always. Restaurant cooks, in particular, were prone to use an axe or a big knife. Deputy Sheriff John Holmes was killed in August 1910, by Riley Wooten, a miner employed on the Stall lease, who shot him with a rifle in front of the National Hotel. He was brought to trial and found guilty of murder. The *National Miner* said Wooten was drunk when he did it, but according to John Pelton, it was a cold blooded killing by a "gang of toughs . . . the whole thing has been brought about by the agitating Western Federation [of Miners] element."

During the year that George Pelton lived in the town he got a heavy dose of the wild and wooly side, although

probably no wilder than he had already known in Rawhide, Wonder and Goldfield. Goldfield, in particular, he
remembered with affection for the rest of his life for all
the beautiful girls, "More beautiful than Hollywood,"
who flocked to the fabulous camp to look for rich hus-
bands. Ugly things happened in Goldfield, too. A man
who sang in Tex Rickard's saloon ("The Road to Man-
dalay" was his specialty) kicked his own dog to death
in the street while no passerby dared to interfere.

In Wonder, rooms were in such short supply that
George shared his with an all-night gambler. The gambler slept there by day and, at night, George would fall
asleep listening to the tinkle of the pianos in the honkey-
tonks. One evening, while he was in a saloon, three of
the customers started shooting. The man standing next
to George was hit, spun around, and fell to the floor.
There was no time for a general stampede from the
room, the way the movies like to show it. George, unhurt,
jumped behind the bar, where there stood a big iron
safe — unlocked. He pulled the safe door wide open and
stood behind it, hoping the heavy steel would stop any
bullets flying his way. The saloon keeper, who saw what
George had done, ran over to get behind the safe door,
too. The door wasn't big enough to cover two men, so
he jerked George away from his impromptu shield and
got behind it himself. The gunplay was still going on so
George, in turn, pulled the man away and again popped
behind the sheltering door. The saloon keeper, red with
fury — and still dodging gunfire — hauled George out
a second time. By then the battle was over, but the man
was so angry and upset by the tug-of-war that he com-
plained later to John Pelton that George wouldn't even
let him hide behind his *own* safe door. When the Peltons
only laughed, it didn't help his disposition.

In Rawhide, George had a friend who was a compulsive gambler. The man would play blackjack for three days and nights at a stretch, and was frequently drunk. George, who neither drank nor gambled, found himself entrusted with the money of his friends who did. At the time it didn't occur to him that walking around with thousands of dollars in a money belt might not be the safest thing to do in a wide open town.

While he lived in National, George's regular companions at a small eating place were a man who had killed twice and a town prostitute who said she was the daughter of a missionary. The man was later sent to prison for murdering a third time, and the girl was killed one night by a gang of four or five drunks who knocked on her door and fired their guns through the wall when she didn't answer. "As usual," George said later, "nothing was done about it."

Not all incidents ended fatally, to be sure, but plenty of them ended in a bloody shambles and even naturally peaceable George Pelton could not go untouched. A man he knew from Rawhide shared George's tent and served as his body guard until, on his own time one night, the man got into a no-holds-barred saloon brawl with two men and bit off the upper lip of one of his attackers. Figuring that his bodyguard was a net liability after that dust-up, George let him go, and found it a relief when he was able to move from the rowdy town over to the National Mine headquarters in Charleston Gulch.

In July 1910, an accident at the Stall brothers' lease caused serious injuries to both George and Frank Stall. They were at the bottom of a 400-foot shaft, sorting bonanza rock after blasting, when the hanging wall caved and the brothers were caught by a three-ton rock fall. John Pelton and a crew from the National Mine scrambled to the rescue, but George Stall was buried for half an hour before a couple of men could dig him free. Both brothers were severely hurt. George Stall had a broken left leg, a fractured right shoulder, and an injured back; Frank's right leg was badly broken in two places below the knee. Their half-brother, William Lehman, was injured less seriously. Two doctors hurried from Winnemucca to treat them, and the Stalls' sisters, Mrs. Enried and Mrs. Fletcher, came from Marysville, California, to nurse and care for them.

Recovery was a long drawn out matter for both brothers, and personal supervision of the work on their lease was out of the question for months. While their lease was producing about $250,000 per month in highgrade ore, according to the *Mining & Scientific Press* of November 1910, they were faced with the prospect of labor troubles and the increased need for pumps and other heavy machinery. Under the circumstances, they accepted an offer from the National Mining Co. to buy out their lease for $150,000 spot cash and a chance to mill the ore already broken. At the end of August the Peltons took possession of the Stall lease and double bulkheaded all connections except through Tunnel 4, where they installed a complete change room.

The Stall lease was not the only one in operation on National Mine ground, but it was the only one whose terms had not been satisfactory to George Pelton. Before he had time to relax at having the Stalls out of the picture, though, the Peltons saw their possession of the mine threatened from a new quarter — a series of lawsuits over the ownership of the gold bearing vein itself. These suits were based on the apex law of the United States, a law so difficult to understand that U.S. Geologist Waldemar Lindgren called it, "A constant and irresistible temptation to clever interpreters of the statutes."

9

The Apex Problem and
The "Great Treasure Vault"

The law of the apex, or the law of extralateral right, was passed by Congress in 1872. It provides, basically, that the owner of a claim wherein a vein of ore reaches its apex, or highest point (often where it intersects with the surface of the ground), is entitled to follow the dip of that vein as far downward as it goes, even if it should dip beneath the surface of a neighboring claim. There are restrictions: the vein must pass through both ends of the claim, and those end lines must be parallel. By law, a lode claim can be no larger than 1500 by 600 feet. The owner of the claim can follow his vein to the side, therefore, but his "share" of the vein is limited by his side lines to a maximum of 1500 feet.

The basis of extralateral right, then, is threefold: one, the apex of the vein must occur within the surface boundaries of the claim; two, it must pass through both end lines of the claim; and three, the miner must be able to prove that the vein he follows downward is the same vein as the one whose apex occurs within his claim lines.

That's where a lot of trouble begins. Veins are not likely to be solid slabs of ore, dipping downward in neat parallel striations. They wrinkle, and "pinch," and change directions, and fork. It is enormously difficult for the locator of a claim to judge the course and strike of a vein from the little bit that is exposed as outcrop. A vein almost never turns out to be as straight and simple as the politicians who passed the apex law thought veins should be.

The apex law has been denounced by experts for over a hundred years. "Marvelous perfection as a piece of trouble-breeding, perplexing and ambiguous legislation," said one judge. Another deplored the fact that even after a number of cases had been decided by the U.S. Supreme Court, the lower courts couldn't decide on the proper construction to give to the Supreme Court's decisions.

The law is still on the books.

The Peltons were mindful of the apex problem. Not long after they started running the mine, they discussed with Scotten and Snydacker the possibilities of buying control of Workman's ground where it might cause apex trouble. They opened negotiations with Workman but on April 24, 1910, John Pelton wrote to Chicago that Workman was asking $150,000 for ninety per cent of the stock in the company whose claims touched their western sidelines. In Pelton's opinion, this was much too much money, since the lawyers advised him that there was no danger concerning extralateral rights. When they needed all the cash they could get to pay for machinery and timber and supplies to develop the mine, $150,000 for apex insurance seemed an unthinkable sum.

Lawyers' assurances were not enough for John Pelton. A few weeks later he wrote to Scotten and Snydacker, "I proved the apex by following the Charleston vein directly up to the surface — making the shaft at the surface extending down to our level 3½ and from 3½ to 4." As a double precaution he compelled the Stalls to start a shaft at the apex on the surface and follow the vein with their shaft to whatever depth they might attain during the lifetime of their lease.

Beyond question, John Pelton proved rule one: the apex of the vein most certainly did occur within the boundaries of the National Mine claims. Rule two wouldn't be so easy. In August 1910, came the first apex suit against the National Mine. According to court records, H. E. Orr, claiming possession of the West Virginia #1, Charleston #1 lode, West Virginia fraction lode, and apexes, under contract from the Charleston Hill National Mining Syndicate, the owners of these claims, said that the Charleston lode on its strike passed through into the West Virginia #1 lode. He claimed trespass and asked the court for a survey. In effect, he was saying that the edge of the vein did not pass through *both* end lines of the National Mine claim, a pre-requisite to the existence of extralateral rights. His West Virginia #1 and Charleston #1 claims shared common side lines with two of the National Mine claims. If the Charleston lode crossed a side line onto one of his claims, then he had a piece of the vein. If the Charleston lode passed through both end lines of the National Mine claim, Orr could not mine it at all, even though it ran beneath the surface of his own location. The Humboldt County judge ordered a survey made of the boundaries and workings of the National Mine property, while stipulating that no action was pending between Orr and the National Mining Co.

The National Mining Co. thereupon petitioned the Superior Court for a writ of certiorari. They said the district judge had exceeded his jurisdiction in ordering the survey and that if Orr wanted to prove trespass he should produce a survey of his own property and not compel a neighbor, in whose property he had no interest, to have to do so. After delays and continuations, the Superior Court rendered judgment on June 20, 1911. The court upheld the petition, saying that since there was no lawsuit pending, the judge had no authority to order a survey of the National Mine.

There was another matter pending at this time. The Peltons and the Chicago investors decided to form a new corporation, to be called the National Mines Company, incorporated in the state of Wyoming. The date of incorporation was September 19, 1910, and Buckley, Scotten and Snydacker all came from Chicago to attend a meeting at the mine. The new corporation was capitalized at one million shares, $1.00 par value per share. In practice, it would appear that the stockholders simply traded their shares in the National Mining Co. for shares in the new National Mines Company. The board of directors remained virtually the same: the two Peltons, Sam Scotten, and L. G. Campbell were re-elected, and J. G. Snydacker replaced Jerry Sheehan. John Pelton was again named president and George continued as secretary.

The group could look with satisfaction on all they had accomplished in the year since they took over the National Mine. They had built roads, run about two miles' worth of tunnels and other workings, put up a mill, tramway, assay building, watch tower, office and living quarters, and to cap it all, had managed to start paying dividends.

The vertical depth of the mine was now below 900 feet. One hundred men were employed, half of them in development work and, rag town or no, the town of National was able to retain a substantial population even with winter coming on. In the election of November 1910, two hundred and twelve registered voters cast their ballots at National. The bullion no longer had to be shipped under guard to Winnemucca. Since September it had been sent to the mint by registered mail, in four pound bars, and the automobile bullion convoys didn't have to buck winter snowdrifts.

LIST OF STOCKHOLDERS

NATIONAL MINES COMPANY, MARCH 23RD, 1911.

..pany Treasury,	240,001
L.G.Campbell, Winnemucca, Nevada,	3,501
Robert Lewers, Reno, Nevada,	500
Fred Grob, " "	2,250
Wm.Levy, " "	1,250
R.S.Bolam, National, "	2,700
Mrs R.S. Bolam, " "	300
John F.Harris, 15 Wall Street, New York City,	3,000
Henry Rogers Winthrop "	1,000
William Blattner, Paradise Valley, Nevada,	500
W.J.Hotchkiss, Fife Building, San Francisco,	26,500
E.B.Deming, South Bellingham, Washington,	25,000
A.W.Deming, " " "	898
C.Y.Deming, " " "	2,000
C.W.Buckley, Chicago, Illinois,	53,686
J.G.Snydacker, " "	154,083
S.C.Scotten, " "	154,083
F.L.Deming, " "	23,000
H.H.Hitchcock, " "	7,000
F.H.Rawson, " "	16,000
C.M.Mitchell, " "	3,750
C.M.Boynton, Medford, Oregon,	1,000
Jno.R.Tomlin, " "	1,000
Wm.T.Gould, Mills Building, San Francisco,	4,100
Geo.S.Pelton, National, Nevada,	101
John E.Pelton, " "	81,251
M.E.Atkinson, Gordon Hall, Mills Building, San Francisco,	500
Mrs.R.Lyance, Chicago, Illinois,	300
A.Hieronymus, " "	5,000
Darius Miller, " "	50,000
Ella F.D.Kennett, " "	10,000
C.E.Wilcox, " "	7,000
Louis Eckstein, " "	22,000
Elsie (S) Eckstein, " "	2,000
J.P.Molloy, " "	23,000
C.H.Sullivan, " "	1,000
Margaret Deming, " "	1,000
E.G.Deming, " "	1,000
Pauline Deming, " "	1,000
Helen E.Deming, " "	1,000
E.S.McCord, Mutual Life Building, Seattle,	4,000
J.E.Gorman, Chicago, Illinois,	1,500
H.Ryan, " "	300
C.V.Indierrieden " "	300
J.S.Indierrieden " "	180
G.E.Kingsbury " "	500
B.F.DeMuth, " "	1,500
L.Benzing, " "	156
Lydia Benzing, " "	60
H.A.Foss, " "	4,000
Frank R.Fardridge," "	1,000
George B.Harris, " "	26,750
Henry Kerchner, Paradise Valley, Nevada,	500
M.L.Goodkind, Chicago, Illinois,	500
S.S.Downer, Reno, Nevada,	3,750
Prince A.Hawkins Reno, "	1,250
Frank C.Letts, Chicago, Illinois.	4,000
Mrs.Frank C.Letts, " "	1,000
Thomas J.Prindiville," "	4,000
John Dupee, " "	2,000
Frank P.Frazier, " "	4,000
W.H.Lehman, National, Nevada,	5,000
John P.Cobb, 24 West 74th, St., New York City,	500
Total Shares	**1,000,000**

The State of Wyoming

OFFICE OF THE

SECRETARY OF STATE

▣——▣

United States of America, ss.
State of Wyoming.

I, Wm. R. Schnitger, Secretary of State, of the State of Wyoming, do hereby certify that the annexed copy of the Certificate of Incorporation of

THE _ NATIONAL _ _ MINES _ _ COMPANY

has been carefully compared with the original, filed in this office on the NINETEENTH day of SEPTEMBER, A. D. 1910, at 9:30 o'clock a. m., and is a full, true and correct copy of the same and of the whole thereof.

In Witness Whereof, I have hereunto set my hand and affixed the Great Seal of the State of Wyoming. Done at Cheyenne, the Capital, this NINETEENTH day of OCTOBER, A. D. 19 10.

Wm R Schnitger

Secretary of State.

By L C Hinkle

Deputy.

A NEW NAME: THE NATIONAL MINES COMPANY, IN 1910
Courtesy of the Secretary of State, Wyoming.

State of Nevada ⎫
County of Humboldt ⎬ S. S.

_____Jno E.Pelton_____ - _____ and
_____George S.Pelton_____, being duly sworn, depose and say that they are
respectively President and Secretary of the _____National Mines Company_____

a corporation organized under the laws of ___the State of Wyoming___
and that the following is a full, true and correct statement of facts required by section 85 of the
Corporation Law of the State of Nevada, as follows, to wit:

NAME OF DIRECTOR	Date of Election	Term of Office	Character of Business	Postoffice Address (street and number, if any)
Jno E.Pelton	1910 Oct.15th	1yr	Mine Manager	National, Nevada.
George S.Pelton	Oct.15th	1yr	Book-keeper	National, Nevada.
S. C. Scotten	Oct.15th	1 yr	Broker	39 Board of Trade, Chicago, Ill
J.G.Snydacker	Oct.15th	1 yr	Broker	39 Board of Trade, Chicago, Ill
L.G.Campbell	Oct.15th	1 yr	Lawyer	Winnemucca, Nevada

Furthermore that___George S.Pelton___ of National
___Street,___ Humboldt, ___County, Nevada,
(City or town)
is the duly and regularly authorized Agent for said Corporation upon whom process can be served.

Jno E. Pelton
President.

Geo. S. Pelton
Secretary.

Subscribed and sworn to before me this _____day of
_____, 1910.

S.H.Cruikshank
Notary Public.

NATIONAL MINES COMPANY OFFICERS, DECEMBER 1910
From the author's collection.

Highgrading remained a problem. In November, bandits stole some sacked highgrade ore from the Tunnel 3 storage area. A month earlier, $25,000 in National highgrade turned up in a hotel in Oakland, California, when the proprietor turned in the thief before he could sell it to the mint.

The Lane mill was giving them trouble. They discovered that they would have to replace the mill foundations with a solid concrete base; dressed stone and concrete pads would not do. Furthermore, clean-ups were frequent and slowed production.

Things were looking up to such a degree that the Western Pacific Railroad hypothecated a line to run north from Winnemucca up the Quinn River Valley.

The *National Miner* brought out a "retrospective" edition on December 9, 1910. The banner headline proclaimed, "THE GREAT TREASURE VAULT OF NORTHERN NEVADA . . . Its Complete History Up To Date." The front page carried a photograph of the National Mines Company mill, captioned, ". . . turning out monthly $250,000 of gold bullion." An effort was made to give space to other properties and leases, but the National Mine was the big story and got all the pictures. One cut showed S. C. Scotten and J. G. Snydacker and another showed John E. Pelton standing next to C. W. Buckley. An inside page headline reads, "Freight; Yes, 100,000 Pounds." Frank J. Frey, head of Frey's Big Mules Outfit, is quoted as saying he has 100,000 pounds of freight "at the foot of the hill." There was mine timber, merchandise, and odds and ends for everyone in camp. In the Local and Personal column, "the Pelton auto" scored as an agent of mercy on its most recent trip to Winnemucca. One passenger was S. K. Bradford going to see his sick wife in California and another was Ed Kellett, seeking medical help for a broken shoulder. The newly elected officers of the miners' union got a couple of column inches; James Trainer was named its new president. There were a lot of personal tidbits about local people which are the lifeblood of a local paper, but nothing about coming Christmas festivities and no advertising aimed specifically at the Christmas market.

The *Miner* that appeared five days after Christmas reported the celebrations in detail. The schoolhouse was decorated with holly and a tree was given by John Pelton. There were exercises by public school and Sunday school classes and organ music by Miss Langwith and Miss Leonora Pelton. Miss Langwith was presented with a silver napkin ring by her pupils and, for the grand finale, Santa Claus stood by the tree and gave each child a Christmas stocking. The National Mines Co. gave everyone the day off with pay.

The early months of 1911 saw fantastic production at the National Mine. The *Miner* quoted "an experienced mining man from Salt Lake" as saying that,

> Half the men working in the National Mines property go armed, and shift bosses carry six-shooters instead of picks . . . over 120 men are employed, and he has been told by men working in the mine that they have taken out 100 pounds of ore in one shift, which went $70 to the pound, or $7,000 for one man's eight hours of labor.
>
> From what I see and reliably hear, this is the greatest mine the world has ever known, and you do not hear much about it on the outside, either . . . I never saw anything like it for rich free gold ore, going from $40 to $100 a pound. If I should tell you just what I have seen you would think me crazy.
>
> That mine is a freak of nature. No one ever saw anything like it before and probably never will

again. The ore is free milling, but runs high in silver. When you can't see free gold in the stope it is dubbed "no good."

President John E. Pelton lately took out one boulder to Frisco, which will average $70 a pound, or $7,700 for the 110-pound chunk . . .

It was clear that the Peltons' early faith in the mine and their perseverance in the face of enormous difficulty was paying off. The *Miner,* which loved to talk about large sums of money, carried the headline, "HAS PAID HALF MILLION IN DIVIDENDS SINCE JULY," and detailed the specifics:

Another dividend of 10 per cent will be presented on Wednesday of this week, March 1, to those who are fortunate enough to own shares in the National Mines company of National, Nevada. This will be the third dividend this year. On January 1 a 15 per cent dividend was paid; on February 1 a 10 per cent dividend was paid, and the third will be mailed out to reach stockholders the first of the month. This makes an aggregate of 35 per cent in three months of 1911. Further than this it is stated on excellent authority that the company expects to pay a similar dividend monthly.

Just think of it! $80,000 for distribution from this one mine representing thirty days' net earnings.

The figures already given seem large, but when it is stated that since July the company has paid out in dividends $495,000, some idea of the magnitude of the production of the National Mines company and what its property represents may be obtained. The production for the past year by lessees and company workings is practically $1,750,000.

The capitalization of this company is one million shares, but only 760,000 shares have been issued. There has been little outside selling of the National shares, for the management has never cared to list its stock. There have been some sales to those who already held stock, however, and the price has been $3.50 bid and $4 a share asked . . .

The death limit to this property cannot be estimated . . . they are actually producing and marketing from $400,000 per month. And this is without taking into consideration the values of the tailings. It is declared by conservative mining men that the tailings, with which the company has not yet had time to deal, are worth more than the ore taken out of many successful mines . . .

It takes statistics a little time to catch up, but that spring, National was acknowledged as ranking third among the gold and silver camps of Nevada, principally due to the production of the National Mine.

10

New Management

Two weeks after the "Half Million in Dividends" story came a bombshell. "PELTONS RESIGN FROM MANAGEMENT OF THE NATIONAL MINES COMPANY," said the black headline of the *Miner* on March 24, 1911. Underneath, a formal photograph of John Pelton was spread over the two middle columns, with a black border around it as if he were dead. A smaller, less funereal photo of George Pelton showed him in his working clothes: trousers tucked into high laced boots, a jacket, vest, and broad brimmed hat. The story details were meager:

John E. Pelton and Geo. S. Pelton, as general manager and secretary of the National Mines company, have tendered their resignations to the board of directors of the company.

The change caused by this action on the part of the Peltons only applies to the practical management of the company at National, or the direct operating management at National will rest on new shoulders.

The successors for the manager and secretary are not designated as yet, but the same will be elected at a regular meeting of the board of directors in the early part of April at National.

The Peltons will retain their stockholdings in the company and John E. Pelton will continue as president of the National Mines company, thus emphasizing the fact that the withdrawal as general manager and secretary is not a result of friction, but with a desire to retire from the vexatious responsibility and work which the great obligation involved in such a position exacts from a fair and conscientious man . . .

The *Miner* went on to discuss the personal histories and admirable characters of both Peltons, but gave no further background on the reasons for their decision to leave the direct management of the mine.

The Peltons' move was no surprise to the other directors. For some time there had been a hardening difference of opinion between the two Peltons and the Chicago group. The Peltons had learned through experience to proceed cautiously in the development of a mine, to invest in material and equipment only when such expense could be well justified by the ore already exposed. The Chicago group, who had the money but not the same hard-won knowledge of mine development, were far more inclined to "think big," to install more elaborate machinery than current production would indicate, betting that future production would expand to meet the enlarged facilities and milling capacity. The Peltons had seen too many mining properties fail to develop as their backers hoped, beautiful but shallow deposits that swallowed up more investment than they ever returned. The National Mine represented the Peltons' best hope of making a big stake. For the Chicago men, their capital was venture capital, not their one and only bright hope. This difference in viewpoint was so basic that the Peltons felt that it was time to cash in their stock when they could no longer persuade the Chicago group to see things their way.

There was also the matter of George Pelton's health. Whether it was due to bad diet or the nervous stress of his position and responsibilities, George had developed a painful and serious stomach ulcer. As soon as arrangements were complete, George left for New York to seek medical attention.

Weeks before they resigned, John Pelton had been in Pasadena, California, dickering for the sale of some of his stock and making arrangements to buy a house. Some shares he sold at four dollars a share, some at three. Early in March, he closed the deal for the house in Pasadena. Of the buyers for his stock, he wrote George that, "They are all satisfied to take the chance of assuming the risk and in making the trade I tried to discourage them, they are the aggressors."

On March 31, 1911, the *National Miner* announced:

THE NEW MANAGEMENT

H. A. Foss, general manager

F. A. Taylor, assistant manager

The new general manager and assistant manager of the National Mines company were elected at the meeting of the board of directors held in camp Wednesday, following the arrival here of S. C. Scotten, treasurer; L. G. Campbell, vice-president, and C. W. Buckley, the latter being one of the heaviest stockholders of the company. Messrs. Foss and Taylor accompanied the party of company officials and stockholders.

The Peltons handed in their resignations to the board of directors at that Wednesday meeting, and the new board was now made up of J. G. Snydacker, president; C. W. Buckley, secretary; H. A. Foss; and S. C. Scotten and L. G. Campbell, hold-over directors.

A smaller headline bade "FAREWELL TO MISS PELTON" and followed with a column of incredibly fulsome prose, complete with a quotation from Cicero. Boiled down, it reported a surprise party held at the Charles Haeffners for about thirty people. Unquestionably, Leonora Pelton was well liked. She had been active in all the local doings and was highly valued for her ability to play the piano. According to the *Miner,* the party was "an explosion of goodness and kindness for the valuable and fair and cherishable association that Miss Pelton always accorded to her friends and the entire community of National while in our midst. There never was an entertainment or public movement for the entertainment and edification of the public that Miss Pelton did not dedicate her talents, ability and art to aid such an occasion."

One of the guests at the party felt a special pang at seeing her ready to go away. P. R. Whyttock, mill superintendent and metallurgist, had fallen in love with Leo Pelton. It wasn't long before he left National himself, and shortly thereafter persuaded the lady to marry him.

The Peltons lost little time in leaving the camp. Moving was a simple matter of packing a few suitcases. The house in Pasadena was ready and waiting, and the John Peltons had always kept an apartment at the "Colonial" in Reno. National never had been home to them. Although they made a lot of money there, they left with a strong sense of resentment. They felt badly cheated, not by the Chicago group whose backing they had needed, but by Workman and his cohorts who put them in a position where they were forced to seek such backing, forced to share control of the mine, and forced to give up a giant share of the profits. Just how big a share is evidenced by the list of stockholders as of March 23, 1911, which shows over 500,000 shares owned by the Chicago financiers and their friends — *more* than the number of shares originally optioned by the Peltons!

The Chicago group were now in full control of the National Mine. If they made mistakes, they could blame no one but themselves.

How were things going to be different under the new management? There was shifting at the top levels, of course, but other than that, not a lot. The Chicago men named their friend H. A. Foss, to be the new general manager, temporarily, until the position was taken over by Frank A. Taylor. Robert S. Bolam, a "Pelton man," stayed on until mid-June, when Perry G. Harrison was named as his replacement. On a lower echelon, Charles Haeffner, the company driver, bought an almost-new Stearns automobile from the company and started his own auto livery business.

The day-to-day routine of mining and its concomitant, highgrading, were little affected by the top level changes.

As spring wore on, George Shea made improvements to his property near the ridge that stood between the mine and the town. It was known as "Sheatown" or "Sheaville," and the *Miner* remarked that, "As a convenient business location his boarding and commercial house is clearly centrally located for the operating and producing section of National." Since every man who lived in camp and worked in Charleston Gulch had to pass by Sheatown going to and coming from work, it surely was. Convenient, indeed. Today, when a miner in Winnemucca remembers Sheatown, he does so with a sly grin, as Shea's location soon became the red-light district.

A curious problem cropped up that summer. The town of National and all the mines asked to be taken out of the Santa Rosa National Forest reserve created on April 1, 1911, by President Taft. The people were growing bitter about permits and dry laws, and got up a petition to send to Congress. The Bureau of Land Management agreed that one section should be taken out of the National Forest, but one section (640 acres) was not enough area so, in October, National was eliminated from the National Forest by presidential proclamation. Including the townsite and the mines, 1,400 acres were thus liberated.

Waldemar Lindgren, a government geologist, described the mining camp that June of 1911 as consisting of, "fifty frame houses and about 100 tents, the National Hotel, a two-story structure, being the most striking object. No trees shade the hills, but, as seen on a June day, the gray-green mantle of the luxuriant sagebrush and the broad splashes of yellow wildflowers, against the dark reddish-brown of the basalt flows of the high ridges, give the color scheme of the landscape." There were about 400 people in National then and about another 450 in McDermitt, down on the valley floor.

Transportation was getting a little easier. Most of the important people, including Jesse Workman and the Stall brothers, had their own automobiles. The best cars of the district were ones like the "Haeffner (ex-National Mines Co.) auto," a red 60-horsepower Stearns with 42-inch wheels and a 14-inch ground clearance. These big cars could make the run to Winnemucca, sometimes, in four to five hours.

Death Valley Scotty made a visit to National. The *Miner* said the purpose of his visit was unknown, but it is likely that both Scotty and National were running true to form. If dark rumors from Goldfield were true, Scotty was more than partial to a little highgrading, but by the time Scotty got to National, the "Gumshoe Kid" and his friends had matters too well organized to let Scotty muscle in.

In their unending battle against this problem, the National Mines Co. put up a fancy new change house with 140 lockers "with Yale locks" and four pools with hot and cold water and "adjustable shower baths." The miners went on a short strike over the change rooms. General Manager Foss was able to smooth matters over and the company had a housewarming and dance to celebrate the event and proclaimed that one day a week would be ladies' day — for the baths, that is.

Fourth of July celebrations at the camp were lovingly reported by the *Miner*. In addition to a brass band and the Goddess of Liberty, the morning parade included the labor unions, the National fire department, and the G.A.R. Mayor Tom Defenbaugh, the councilmen, the city treasurer, and Jesse L. Workman all rode in carriages. Frank Frey's large freight outfit was loaded with bullion from the mines and there was a maze of bunting, festoons of streamers, and flags of all nations.

After lunch came the speeches, a reading of the Declaration of Independence, and the athletic events. There was something for everybody: rope jumping, sack races, fat man's race, ladies' race, and on and on. The evening finished off with a band concert, fireworks, and "dancing on the Gretna Green."

The summer progressed placidly. The four-pound bars of bullion kept traveling from the National Mine to the Mint via registered mail. The *Miner* estimated that dividends paid out by the National Mines Co. in one year came to about one million dollars. Profits like that encouraged development work. William J. Bohrman (a suitor of C. W. Buckley's elder daughter) and mining engineer Perry Harrison took a six-mule outfit up Martin Creek, over the mountain to the east of Charleston Hill, surveying water courses and establishing water rights for water needed to treat the ores of National. H. R. Brown, mill superintendent and metallurgist who succeeded P. R. Whyttock, planned to install a stamp mill at the National Mine. At other mines in the district, engines and a blower were going in at the Cheefoo claim on Auto Hill, and the Hyde-Prout lease on the Fairview claim found good ore.

The whole district believed it was just a matter of time before another property would develop as richly as the National Mine.

11

The Battle of the Apex

In August, the Mammoth National Mine struck high-grade.

The Mammoth Tunnels 1, 2, and 3, entered the mountain from the Charleston No. 1 claim and the West Virginia No. 1 claim. These claims belonged to the Charleston Hill National Mining Syndicate and were leased to the Mammoth Mines Company, operators of the Mammoth National Mine. Unfortunately for Mammoth, the Charleston No. 1 and the West Virginia No. 1 adjoined the Charleston and West Virginia claims of the National Mines property. Immediately, the National Mines Co. asked for an injunction, based on the apex law.

To try to make it clear: the West Virginia claim belonged to the National Mines Co., while the West Virginia No. 1 did *not*, and the Charleston claim belonged to the National Mines Co., while the Charleston No. 1 did *not*. The claims with "No. 1" in the name were to the west of the ones with plain names. The plain names belonged to the National Mines Co., and "No. 1's" belonged to the Mammoth National.

The injunction sought by the National Mines Co. was granted, restraining Mammoth National from further operations in that part of the mine where the highgrade was found.

In November, to confuse matters worse, Jesse Workman, head of the Charleston Hill National Mining Syndicate, appeared and demanded possession of the

Mammoth Mine, declaring that the Mammoth National had defaulted in its agreement with the Workman interests. Mrs. Workman accompanied her husband and was present when Workman closed up Mammoth Tunnel 3 with a lock and key, surrounded it with armed guards, and posted a notice forbidding anyone to enter without his permission. The manager of the Mammoth National, a Mr. Baxter, gained access to the mine through Tunnel 2 and there he settled in, prepared to endure a siege. The dispute, which wasn't settled until after Christmas, was concerned only with ownership interest in the Mammoth National. The apex litigation with the National Mines Co. remained a separate and much more dangerous issue.

During the winter the apex suit slowed everything down. Neither the National Mine nor the Mammoth could mine the vein in dispute, and the production standstill affected the whole life of the camp. Even the ever-present highgrading was noticeably reduced. Frank Taylor, the National Mines manager, turned his energies to re-working the mill to save more of the tailings and lower grade ores. Then cold weather froze up the mill. He had to rent a couple of autos from Haviland and Hoskins, load them to capacity with highgrade, and send them down to Winnemucca. There were thirteen sacks, average weight 145 pounds, filled with ore running $30 to $85 per pound, so the cargo of highgrade *untreated ore* carried a value of $55,000 to $100,000.

Even though the mine was pretty well shut down, the searchlights continued to play all night to prevent dump poachers and highgraders from making their getaway. Armed men stood guard day and night in the bonanza shoots, and the foremen wore guns.

Usually, the mule teams hauled down only rich tailings and concentrates for shipment to the Selby smelter. The tailings, which ran $150 a ton and up, were left after the higher grade ores were milled and their values recovered as bullion which was poured into small bars which went to the Mint by mail. Tailings of such high value indicate an inefficient milling operation, but ore as rich as the National highgrade is very forgiving to the profit and loss statement. It is standard practice at all mines and mills to keep a tailings pile for later re-working, and this was done at National. In fact, as time went on, the location of the tailings piles presented a problem in the narrow gulch at the foot of Charleston Hill.

The spring thaw finally came, melting the snow and un-freezing the mill. Various auto lines brought back the miners who had "gone south" for the winter. Daily auto service meant daily mail, too. In May, Haviland and Hoskins discontinued their horse-drawn stage line. Either the roads, or the cars, or the weather, were getting better. Maybe all three.

The re-vamped mill began reducing concentrates from the Stall dump, which the "big" mule teams freighted down to Winnemucca at 20,000 pounds per trip. As the volume of freight mounted, the fourteen-horse team of the Verdi Lumber Company which had been hauling between Elko and Tuscarora was now put on the Winnemucca-National run.

All this commercial activity contrasted with the pall of idleness in the two major mines. The pending lawsuit over apex rights meant that ore reserves could be developed and blocked, but no one could touch the ore bodies at question in the apex suit which was due to open in Carson City on June 17, 1912, in the U.S. District Court, Judge E. S. Farrington presiding.

As time for the trial drew near, Carson City was crowded with Nationalites. John Pelton was on hand, regaling friends with the tale of his adventures in Mexico the previous winter, when he and three other men were held up on the Balsas River in the state of Guerrero by forces of the revolutionary bandit, Jesus Salgado. Their guns were taken but they were otherwise well treated. George Pelton did not come to the trial. He had gone East when he left the company to seek treatment for severe stomach ulcers, and did not yet feel well enough to travel. Surprisingly, Sam Scotten appears to be the only one of the Chicago backers to come to the trial.

GREAT BATTLE TO DETERMINE
APEX OF THE BONANZA VEIN
NOW ON IN FEDERAL COURT

boomed the *National Miner* as the trial began. In essence, the National Mines Company alleged trespass against the Charleston Hill Mining Syndicate and the Mammoth Mines Company. At this point, Mammoth no longer existed as a separate entity, having been absorbed into the Syndicate following the Workman barricade. Waldemar Lindgren (Bulletin 601, U.S. Geological Survey, 1915) boiled down the question at law:

> The complainant asserted that the apex of the valuable vein crossed both end lines of the West Virginia claim, and that therefore extralateral rights existed, which conferred the privilege of following the vein on the dip outside of the side lines of the claim in which the vein apexed.
>
> On the other side, the defendants, who had been driving the Mammoth tunnels Nos. 1, 2, and 3 close to the side line separating the West Virginia from the West Virginia claim No. 1, claimed that only

The 1912 Apex Lawsuit Map of the National and Mammoth Mines

CHARLESTON

No 5 Tunnel

CHARLESTON No 1

WEST VIRGINIA

MAP
OF THE
MAMMOTH and NATIONAL MINES
Humboldt County
NEVADA
Scale 527 ft = 1 inch
1912

CHARLESTON FRACTION

WEST VIRGINIA No 1

Air Shaft

Caretaker Shaft

Shaft

WEST VIRGINIA FRACTION

Gaudin Shaft

a part of the National vein existed in the West Virginia claim; that where a certain bend or turn exists . . . the National vein had been cut off by an important fault fissure, trending about N. 25° w., which was traced from the gap at the head of Charleston Gulch by the McDonald tunnel, the Caustin shaft, and the levels driven in the same mine workings south of the above-mentioned turn or bend; that the latter turn or bend, in fact, represented places where the later fault fissure intersected the older gold-bearing vein, and that therefore the apex of the National vein did not reach the southern end line of the claim, but that this southern part of the vein was thrown an unknown distance to the northward, and that consequently the National Mines Co. had no extralateral rights as far as the National vein was concerned.

The *Miner* estimated the potential worth of the gold bearing vein to be:

. . . something like $25,000,000, a stake worthwhile for such financial Titans who are embroiled in the contest for such expectant millions.

On one side is a syndicate of Chicago millionaires headed by such men as Sam C. Scotten, A. J. Snydacker, J. P. Malloy, C. W. Buckley, John E. Pelton, R. C. Harris, head of the executive board of the Hill railroad system, and Darius Miller, president of the Burlington Road. Present and directing the aim of these giants of finance is S. C. Scotten.

Opposed to these forces . . . are Jesse L. Workman, founder of National; W. S. Tevis, Al Stanford, of the Borax Smith interests; John Spreckles, W. L. McGuire, Gus Eisen, Clarence Berry and Gordon Campbell . . .

The legal talent comprises the best mining attorneys of the West. On the side of the National Mines

are Hon. Geo. A. Bartlett, Geo. A. Thatcher, L. G.
Campbell (attorney for the company since its earliest days), Judge Dixon [Dickson] and A. C. Ellis,
noted Utah apex and mining lawyers. On the Syndicate side are Messrs. Metson, Campbell, Brown
and Drew, the strongest legal firm of San Francisco;
R. B. Thayer, whose specialty is mining practice,
and Attorney Colby of Lindley & Colby, who are
authorities quoted world wide on mining.

The scientific and expert forces present have been
ransacked from the best that the mining science
affords . . .

This was true. H. V. Winchell, one of the outstanding geologists of his time, headed the cadre that would
testify in favor of the National Mines Company position.
On the other side were an equally impressive group of
experts, including Professor Andrew C. Lawson, head
of the University of California's geology department.

As soon as the trial opened, Mammoth's attorney,
Metson, moved to have the case tried by a jury. He was
overruled. The case would be tried in equity.

Early in the trial, there was some eye-popping testi-
mony by Frank Taylor, general manager for the National
Mine. In describing some of the massive gold deposits
found in the mine, he told of taking out one four-inch
thick plate measuring three by seven feet, valued at
$94,000. Another, three by five feet and an average
of three inches thick, was estimated to be worth $70,000
in its natural state. George Stall, sufficiently recovered
to be able to testify, declared that his lease had shipped
at least one ton of ore that brought a net profit of
$135,000. Two of the shift bosses working in the high-
grade area of the mine, stated that they had each, on
at least one occasion, taken out ore worth over $100,000

per eight-hour shift. Mining engineer Harrison described lowgrade ore as running, "from $35 to $8,000 per ton and considerable of the $8,000 per ton ore." By comparison, at most mines, $50 to $100 per ton ore was considered highgrade. Based on testimony given during the trial, National Mine production was estimated at $6,000,000 including one million in highgrade in thirty months' operations.

These descriptions of breathtaking riches were immediately seized upon and publicized by stock promoters of nearby claims.

The problem, though, was not to prove that the mine was rich, but to establish just who owned the ore. The Charleston Hill/Mammoth group did their best to prove that a fault vein cut the National Mine vein on the bias and displaced it so as to put the southern portion of the highgrade National Mine vein within the side lines of the West Virginia No. 1 claim, owned by Workman et al. They emphasized strongly that Jesse Workman had made the first location on the West Virginia No. 1. This was true, but that first claim was not where the gold-bearing vein apexed. As the days went by, even impartial observers began to feel that the National Mines Company was proving its case on every point.

The trial ran for more than five weeks and cost over $100,000. As matters drew to an end, the question rose of having the Court visit the mines in litigation. The attorney for the Charleston Hill group made the serious charge that such a visit would be to no avail because the physical evidence on which the case was based had been subsequently altered by miners working for the National Mines Co. The plaintiffs hotly resented this charge, and it did the syndicate no good. When a National Mine foreman was called to the stand, he testified

that any work that they had done destroyed none of the
appearance on which the trial was based. The Court
decided to visit the mine workings, accompanied by
experts on each side, as soon as closing arguments had
been heard.

In November 1912, Judge Farrington handed down
a decision in favor of the National Mines Co. The decision goes into precise detail, covering the material
presented at the trial, and concludes, "I am unable to
discover wherein the National vein in No. 4 and Mammoth 3 east, south of the break, differs materially from
the so-called fault fissure, either as to strike, dip, filling
or mineralization. I therefore find that the fault and
the vein are parts of one and the same fissure. The dis-
tinctions which have been called to my attention are
due to causes, forces, and conditions which have been
present at one end of the fissure, but not at the other,
or which have differed in efficiency as they were applied
north or south of the bend.

"Let a decree be entered in favor of the complainant."

The National Mines Company had won.

The successful conclusion of the apex suit did not
mean the end of litigation, however. (In National, there
WAS no end to litigation.) The next step for the National Mines Company was a suit against the Charleston
Hill Syndicate for an accounting to show what profits
the Syndicate had made from development work on
what was now established as the National vein.

Mopping up operations were soon over. On February
10, 1913, a deed was filed transferring the West Virginia
No. 1 and the Charleston No. 1 mining claims (the Mammoth property) from the Charleston Hill National Mining Syndicate to the National Mines Company.

12

Pinched Out

Although the business of mining at the National Mine was virtually in suspension for over a year, there had been some noteworthy developments in the district. The National Nevada Mining Co., Ltd., on Eight-Mile Creek proposed a tunnel project to drain all the mines and produce a lot of water. A dry spell in August of 1912 threatened the National mill operations, and the tunnel propect aroused a lot of interest.

In September the National Mine suffered a fire in which the office-superintendent's residence was destroyed. The flames were prevented from spreading to the mill and concentrator. As soon as the fire damage was cleaned up, work began on a new office building and home for mine manager Frank Taylor and his family, as well as grading and excavation for a big engine and hoisting plant that would facilitate going deep.

After more than a year's enforced idleness, the National Mine did not resume with a rush. Judge Farrington's decision came in November and the mill, which had run all summer on an accumulation of concentrates, was not too efficient at best and in the winter it operated in fits and starts. The rich production figures of 1910 had been reduced almost by half in 1911 because of the apex suit, and the figures for 1912 were less still. These lower figures were to be expected while the litigation was in progress, and the owners were entitled to feel that now they could settle back, secure in their ownership of a fabulously rich mine, and just let nature take its course — mining the precious electrum and declaring an unending stream of dividends. The highgraders were equally optimistic.

How dismaying, then, to see production for 1913 sag to a fifth of the previous year! Mill figures were more shocking still: 4400 tons of ore were treated in 1912, but in 1913 the mill treated only 47 tons.

The whole district made enormous efforts to get things going again. In March 1913, the Hatch and Stall lease on the Mayflower announced a plan to install heavy machinery and sink a shaft; the Charleston Hill Gold Mining Company, directly south of the National Mine, purchased a complete plant with a 60-horsepower engine and compressor and power drills to sink a shaft 800 feet; the National Mammoth Extension Mines put a large crew of men to drive the White Rock Tunnel through to the company lease on the First National Mining Company property. Better yet, teams were loaded with $15-$250-per-ton silver ores from the Walker Cheefoo lease on Auto Hill. It was really good news at last that the district had more than one ore shipper.

In June, U.S. Mineral inspectors came to examine all important mining properties of the district that were applying for patents. The National Mine was at the top of the list; others in the group included the First National property, the Charleston Hill Gold Mining Company ground, and the Charleston Hill Development and Mining Company property.

All this activity meant that men were putting more money into the mines than they were taking out — a pitfall the Peltons had long since learned to fear, and the reason that they sold out when the Chicago group became too expansive.

The National Mine still sparked the fever. In May,

a large body of highgrade was struck in Tunnel 5; the new highgrade ran about 2/3 gold and 1/3 silver, a somewhat different proportion than before. Tunnels 6 and 7 were run, deeper down the hill. But the summer of 1913 was the last hurrah. With mine production dwindling, some of the lessees ran up bills for merchandise that they couldn't pay. The vitality of the camp was ebbing away, and the easy money was just a memory. "All prospecting has failed to disclose ore in paying quantities at any other spot" other than the National Mine mourned the *National Miner,* and ended its publication in September.

For the next year, a full force of men was kept working in the National Mine. There was plenty of $100 per ton ore and plenty of highgrade, too. In the summer, a rich shoot was opened up in the deepest working, where mainly silver ore was found.

Even though the production of gold fell steadily, the mill treated 2763 tons of ore in 1914. In 1915, gold and silver production both rose a little, and the mill tonnage rose a lot, to over 18,600 tons of ore. Even so, there was no disguising the fact that National had fallen on hard times. For lack of funds, the public school closed in April 1915. In the same month, the government cancelled the mail contract; the government had to pay shipping charges of $2.25 per 100 pounds parcel post carried between Winnemucca and the post office at National, but the postal rate was only $1.08 and since the local freighting outfits charged $1.50 to $2.00 per 100 pounds, everybody bundled up their freight in fifty pound packages and shipped parcel post. The threatened deficit was enormous. (A new contract for mail was negotiated the next month.) The post office remained in National through 1919 and then was moved to McDermitt.

In 1916, the mine operated at a loss. What happened?

"The vein pinched out," said C. W. Buckley, looking a little pinched himself.

The directors tried to salvage something of the mine by leasing it in the summer of 1917 to the National Leasing Company. The *Silver State News* in Winnemucca got quite chatty about the whole operation, reporting that National Leasing would try to hit the vein at the 1300 foot level; that the entire mine had been leased; that rental was for a minimum of ten months at $2,500 per month; that the underground workings were valued at $200,000 (cost); that the National Mines Company retained the right of way and that National Leasing would handle any National ore from below level Number 5 at cost — all this to let National Leasing get at their deep prospects through, and under, the National Mine.

The selling price of National Leasing stock was followed even more closely than the actual work. During August it fluctuated between 12¢ and 30¢. In September a serious cave-in on level Number Eight forced the lessees to pull back to level Number Five. Luckily, they hit a $10,000 pocket of highgrade ore, some of it the $75-a-pound kind. By October, the price of National Leasing stock sank to 6¢, and for the next four months they found no more pay ore. In February 1918, the lessees suspended work and defaulted on their payments.

In April 1918, the Nevada mining and business activity survey failed to mention National, and the 100-ton mill was dismantled in 1921.

Even after the National Leasing operations turned out to be a get-rich-quick promotional scheme, the idea of leasing still appealed to the mine owners, and in the following years they made other leases, mostly to small

THE SUPERINTENDENT'S OFFICE BUILDING IN 1946
Courtesy of Nevada Historical Society.

THE TOWNSITE OF NATIONAL IN 1947
Courtesy of Nevada Historical Society.

GEORGE S. PELTON AT THE NATIONAL MINE IN JUNE 1955

NEAR TUNNEL NO. 5 IN 1970, THIS MUCH WAS LEFT. EVEN LESS REMAINS TODAY.
From the author's collection.

operators. Running in the red more than half the time, production limped along until 1924, when title to the six claims passed to one Rudolph Winkler. The state of Nevada revoked the corporate charter at that point. Wyoming, under whose laws the corporation was created, waited until 1935 to follow suit.

The mine was still not given up completely. Revaluation of gold in 1933 from $20 to $35 per ounce gave a lot of dormant gold properties a new lease on life. Patrick H. O'Neil acquired the National claims at a foreclosure sale in 1928, and during the 1930's the mine was worked under the supervision of a man named McDonald. Conditions were much more difficult than in the days when George Pelton ran things. According to one man who worked there during the Depression, it was more of a scavenging operation than a mining enterprise. Gone were the compressed air drills, the forced ventilation, the mine cars that ran on steel rails. The men were obliged to manhandle timbers, ore tools, everything. The air was foul and the acid water ate at their clothes and their bodies. Only the highgrading went on undiminished.

By this time, the original camp of National was a ghost. The miners lived in a boarding house at the mine and one or two families lived in small houses at Sheatown.

Other operators hung on by their fingernails, especially the Charleston Hill National Mines Company and the Buckskin National over on Buckskin Peak. The Depression and the revaluation of gold encouraged this but they, too, were forced to fold in a few years. National was ready to call it quits even before the War Production Board closed all the gold mines in the United States in 1943.

When the National Mine closed down, it had ten adits (entrances) 150 to 2500 feet long; three shafts, and seven miles of underground workings. In good times, two electric hoists and an aerial tram had been installed, as well as a 100-ton amalgamation mill and concentrator. An estimated minimum figure of $6,000,000 in precious metals had been produced, at the old figure of $20.62 per ounce of gold, from ore *averaging* $20 to $30 per pound. Some estimates of the total production of the district say it ran as high at $14 million dollars, most of it from the National Mine.

The National was the only mine in the entire district to return money on the investment. There were no other winners. When Lindgren, the government geologist, came to report on the district, he remarked disparagingly that the National veins were, "essentially silver veins of very moderate tenor, and the gold shoot of the National vein is a unique occurrence in the camp." So the gold was all a fluke — National was basically a low grade silver deposit!

Naturally, the operators of other claims couldn't accept that. Such riches as the National Mine produced *had* to be spread around a little! They bought machinery, drove tunnels, built roads and strung power lines. They made National a roaring camp for a few years, but no second lode was ever discovered and hope — and money — could not hold out.

Does Charleston Hill conceal further wonderful gold shoots? Waldemar Lindgren argues against it, as have other experts, but dreams die hard. Although it has been sixty-five years since the gold pinched out and the camp died, the district has retained enough attraction for someone to keep title to the ground. The present owners of the National Mine, Ralph A. and William E.

Whelchel, bought it in 1949. The Whelchel brothers are as reluctant as their predecessors to accept Waldemar Lindgren's diagnosis. William Whelchel cites correspondence with Fred Searls, Jr., at one time chairman of Newmont Mining Corporation, who as a young geologist had been one of the experts testifying in the famous apex suit. (At the time, Searls was assistant to Professor Andrew Lawson of the University of California, Berkeley. Lawson headed the group of expert geologists who challenged the position of the National Mines Company.) Searls indicated his disagreement with the findings of the court in the National case, saying that he personally believed that the main vein was offset by a fault and does, in fact, continue — somewhere — in Charleston Hill. The Whelchels dream of someday relocating the wonderful vein.

For ghost town fanciers, National is a disappointment. Unlike a Hollywood production, there is no row of dried out buildings straggling down a main street, through which the wind blows with appropriate creaky noises. The tent houses were easy to take down when the people left, and any trace of them soon disappeared. A number of the frame buildings were uprooted and moved piecemeal to other spots such as McDermitt or the camp on Buckskin Peak. Some that stood empty were "highgraded" to repair others. Some were lost to fire. The few remaining ones slowly sagged and rotted until they melted right into the ground. Except for one decaying cabin, the town might never have existed.

The mines themselves are another matter. Tailings and spoil heaps don't disappear so easily, and up and down Charleston Gulch there are some impressive dumps. A couple of them still retain their weathered headframes and ore chutes. The National Mine has a collection of small ruined buildings clustered around the mouth of Tunnel 5. One of them bears a faded sign that says:

NATIONAL MINE
ELECTRUM EXPLORATION
BY
WHELCHEL MINES CO.

The mill machinery is long gone, as is the original mill building. Grazing cattle traverse the hillside, using the old mine paths where convenient, and making their own tracks across the dumps in their way. Rust colored water seeps from Tunnel 5, from whose dilapitated portal some mine tracks still emerge.

The roads, and wagon tracks, and footpaths are fast disappearing. The passing years eat them away and the sagebrush covers them over.

For one glorious year the National Mine was the leading gold producer of northern Nevada and, as long as the rich ore lasted, the whole district was caught up in a frenzy of greed, corruption, litigation, and outlawry. But Fortune didn't linger and National has been obscured by time and eclipsed by bigger, longer-lived rushes. Only the tales of golden riches and of flagrant ore-stealing keep alive the memory of one of the wildest highgrade camps that ever lived.

Bibliography
and
Index

Bibliography

BOOKS

Florin, Lambert. *Ghost Town Album.* Seattle, Superior Pub. Co., 1962

Folkes, J. G. *Nevada's Newspapers: A Bibliography. A Compilation of Nevada History, 1854-1964.* Nevada Studies in History and Political Science, No. 6. Carson City, Univ. of Nev. Press, 1964

Frickstad, W. N. and E. W. Thrall. *A Century of Nevada Post Offices, 1852-1957.* Oakland, CA, Philatelic Research Soc., 1958

Lindgren, Waldemar. *Geology and Mineral Deposits of the National Mining District, Nevada.* U.S. Geol. Survey, Bul. 601. Wash., D.C., 1915

Murbarger, Nell. *Ghosts of the Glory Trail.* Palm Desert, CA, Desert Printers Inc., 1956

Peele, Robert. *Mining Engineers' Handbook.* Second edn. New York, John Wiley & Sons, 1927

Ransome, F. L. *Notes on Some Mining Districts in Humboldt County, Nev.* U.S. Geol. Survey, Bulletin 414. Wash., D.C., 1909

Siringo, Charles A. *Riata and Spurs.* First edn. Boston and N.Y., Houghton Mifflin Co., 1927

Vanderburg, W. O. *Reconnaissance of Mining Districts in Humboldt County, Nevada.* U.S. Bur. of Mines, Inf. Circ. 6995. Wash., D.C., 1938

Von Bernewitz, M. W. *Handbook for Prospectors and Operators of Small Mines.* Fourth edn. New York, McGraw Hill Book Co., 1943

Weed, Walter H. *The Mines Handbook, Vol. XV.* Tuckahoe, NY, 1922

Willden, Ronald. *Geology and Mineral Deposits of Humboldt County, Nevada.* Nev. Bur. of Mines, Bul. 59. Carson City, NV, 1964

Work Projects Administration, Nevada. *Nevada: A Guide to the Silver State.* American Guide Series. Portland, OR, Binfords, 1940

PERIODICAL ARTICLES

Cutler, H. C. "National, Nevada," in *Mining and Scientific Press,* Nov. 5, 1910. San Francisco, CA

Weight, Harold O. "The Ore Merchants of National," in *Westways,* Feb., 1965. Los Angeles, CA

Whelchel, W. E. "Lost Bonanza of the Santa Rosas," in *True West,* Nov. and Dec., 1962. Austin, TX

Winchell, A. N. "Geology of the National Mining District, Nevada," in *Mining and Scientific Press,* Nov. 23, 1912. San Francisco, CA

NEWSPAPERS

Humboldt Star, Winnemucca, NV. Various issues, 1909-1915

National Miner, National, NV. Various issues, 1910-1913

Silver State News, Winnemucca, NV. Various issues, 1909-1915

Index